PRACTICE TESTS

for

PHYSICAL SETTING REGENTS PHYSICS

Published by
TOPICAL REVIEW BOOK COMPANY
P. O. Box 328
Onsted, MI 49265-0328
E-mail: topicalrbc@aol.com • Website: topicalrbc.com

STUDENTS

e purpose of this book is to give you an aid to review for the Physical Setting/ ysics Regents exam, or your school physics exam. You will find answers and planations to the questions from four previous exams. Take your time in going ough each exam. Try to answer each question on your own before checking the swer and the accompanying explanation. Concentrate on those that you have uble with. Do not wait until the last minute to start your review. Start well before : Regents exam and do 20 to 25 questions at a sitting. By the time you finish the ur exams in this booklet, you should have a good understanding of the wording d types of questions to expect on the Regents, or school physics exam.

Good luck on the exam.

Science Teachers

Answers Written By:
Ronald J. Pasto
Owego Free Academy
(Physics —Retired)
and
William Docekal
Science Teacher – Retired

PHYSICAL SETTING REGENTS PHYSICS

Published by
TOPICAL REVIEW BOOK COMPANY
P. O. Box 328
Onsted, MI 49265-0328
1-800-847-0854

June 2008
Part A
Answer all questions in this part.
Directions (1–35): For *each* statement or question, write in the space provided the *number* of the word or expression that, of those given, best completes the statement or answers the question.

1. The speedometer in a car does *not* measure the car's velocity because velocity is a
(1) vector quantity and has a direction associated with it
(2) vector quantity and does not have a direction associated with it
(3) scalar quantity and has a direction associated with it
(4) scalar quantity and does not have a direction associated with it   1 _____

2. A projectile launched at an angle of 45° above the horizontal travels through the air. Compared to the projectile's theoretical path with no air friction, the actual trajectory of the projectile with air friction is
(1) lower and shorter   (3) higher and shorter
(2) lower and longer   (4) higher and longer   2 _____

3. Cart *A* has a mass of 2 kilograms and a speed of 3 meters per second. Cart *B* has a mass of 3 kilograms and a speed of 2 meters per second. Compared to the inertia and magnitude of momentum of cart *A*, cart *B* has
(1) the same inertia and a smaller magnitude of momentum
(2) the same inertia and the same magnitude of momentum
(3) greater inertia and a smaller magnitude of momentum
(4) greater inertia and the same magnitude of momentum   3 _____

4. Approximately how much time does it take light to travel from the Sun to Earth?
(1) 2.00×10^{-3} s   (3) 5.00×10^2 s
(2) 1.28×10^0 s   (4) 4.50×10^{19} s   4 _____

5. A rock falls from rest a vertical distance of 0.72 meter to the surface of a planet in 0.63 second. The magnitude of the acceleration due to gravity on the planet is
(1) 1.1 m/s^2 (3) 3.6 m/s^2
(2) 2.3 m/s^2 (4) 9.8 m/s^2 5 _____

6. Two stones, A and B, are thrown horizontally from the top of a cliff. Stone A has an initial speed of 15 meters per second and stone B has an initial speed of 30. meters per second. Compared to the time it takes stone A to reach the ground, the time it takes stone B to reach the ground is
(1) the same
(3) half as great
(2) twice as great
(4) four times as great
6_____

7. The speed of an object undergoing constant acceleration increases from 8.0 meters per second to 16.0 meters per second in 10. seconds. How far does the object travel during the 10. seconds?
(1) 3.6×10^2 m
(3) 1.2×10^2 m
(2) 1.6×10^2 m
(4) 8.0×10^1 m
7_____

8. A 1200-kilogram space vehicle travels at 4.8 meters per second along the level surface of Mars. If the magnitude of the gravitational field strength on the surface of Mars is 3.7 newtons per kilogram, the magnitude of the normal force acting on the vehicle is
(1) 320 N   (2) 930 N   (3) 4400 N   (4) 5800 N   8_____

9. Which diagram represents a box in equilibrium?

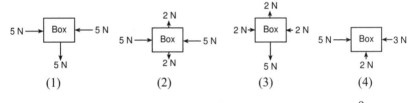

(1)   (2)   (3)   (4)
9 _____

10. The accompanying diagram shows an object moving counterclockwise around a horizontal, circular track. Which diagram represents the direction of both the object's velocity and the centripetal force acting on the object when it is in the position shown?

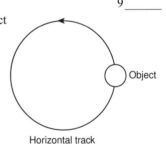

(1)   (2)   (3)   (4)
10_____

11. An airplane flies with a velocity of 750. kilometers per hour, 30.0° south of east. What is the magnitude of the eastward component of the plane's velocity?

(1) 866 km/h   (2) 650. km/h   (3) 433 km/h   (4) 375 km/h   11 _____

12. An 80-kilogram skier slides on waxed skis along a horizontal surface of snow at constant velocity while pushing with his poles. What is the horizontal component of the force pushing him forward?

(1) 0.05 N   (2) 0.4 N   (3) 40 N   (4) 4 N   12 _____

13. A 1750-kilogram car travels at a constant speed of 15.0 meters per second around a horizontal, circular track with a radius of 45.0 meters. The magnitude of the centripetal force acting on the car is

(1) 5.00 N   (2) 583 N   (3) 8750 N   (4) 3.94×10^5 N   13 _____

14. A 0.45-kilogram football traveling at a speed of 22 meters per second is caught by an 84-kilogram stationary receiver. If the football comes to rest in the receiver's arms, the magnitude of the impulse imparted to the receiver by the ball is

(1) 1800 N•s   (2) 9.9 N•s   (3) 4.4 N•s   (4) 3.8 N•s   14 _____

Note that question 15 has only three choices.

15. A carpenter hits a nail with a hammer. Compared to the magnitude of the force the hammer exerts on the nail, the magnitude of the force the nail exerts on the hammer during contact is

(1) less   (2) greater   (3) the same   15 _____

16. As a meteor moves from a distance of 16 Earth radii to a distance of 2 Earth radii from the center of Earth, the magnitude of the gravitational force between the meteor and Earth becomes

(1) $\frac{1}{8}$ as great   (3) 64 times as great

(2) 8 times as great   (4) 4 times as great   16 _____

17. A 60.-kilogram student climbs a ladder a vertical distance of 4.0 meters in 8.0 seconds. Approximately how much total work is done against gravity by the student during the climb?

(1) 2.4×10^3 J   (2) 2.9×10^2 J   (3) 2.4×10^2 J   (4) 3.0×10^1 J   17 _____

18. What is the maximum amount of work that a 6000.-watt motor can do in 10. seconds?

(1) 6.0×10^1 J   (2) 6.0×10^2 J   (3) 6.0×10^3 J   (4) 6.0×10^4 J   18 _____

19. A car travels at constant speed v up a hill from point A to point B, as shown in the accompanying diagram. As the car travels from A to B, its gravitational potential energy

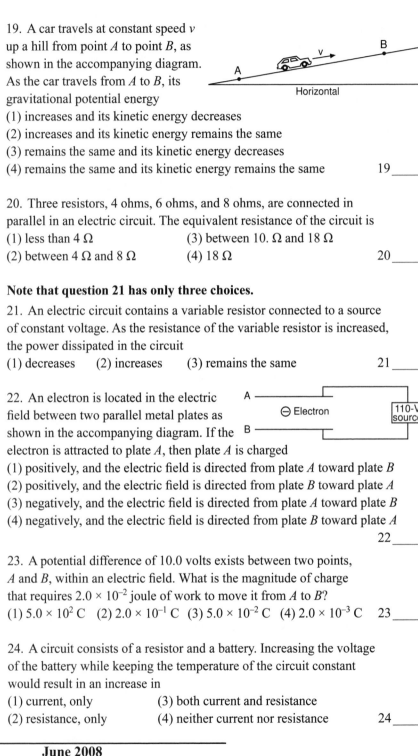

(1) increases and its kinetic energy decreases
(2) increases and its kinetic energy remains the same
(3) remains the same and its kinetic energy decreases
(4) remains the same and its kinetic energy remains the same   19 _____

20. Three resistors, 4 ohms, 6 ohms, and 8 ohms, are connected in parallel in an electric circuit. The equivalent resistance of the circuit is
(1) less than 4 Ω   (3) between 10. Ω and 18 Ω
(2) between 4 Ω and 8 Ω   (4) 18 Ω   20 _____

Note that question 21 has only three choices.

21. An electric circuit contains a variable resistor connected to a source of constant voltage. As the resistance of the variable resistor is increased, the power dissipated in the circuit
(1) decreases   (2) increases   (3) remains the same   21 _____

22. An electron is located in the electric field between two parallel metal plates as shown in the accompanying diagram. If the electron is attracted to plate A, then plate A is charged
(1) positively, and the electric field is directed from plate A toward plate B
(2) positively, and the electric field is directed from plate B toward plate A
(3) negatively, and the electric field is directed from plate A toward plate B
(4) negatively, and the electric field is directed from plate B toward plate A
                                                                       22 _____

23. A potential difference of 10.0 volts exists between two points, A and B, within an electric field. What is the magnitude of charge that requires 2.0×10^{-2} joule of work to move it from A to B?
(1) 5.0×10^2 C   (2) 2.0×10^{-1} C   (3) 5.0×10^{-2} C   (4) 2.0×10^{-3} C   23 _____

24. A circuit consists of a resistor and a battery. Increasing the voltage of the battery while keeping the temperature of the circuit constant would result in an increase in
(1) current, only   (3) both current and resistance
(2) resistance, only   (4) neither current nor resistance   24 _____

25. The time required for a wave to complete one full cycle is called the wave's

(1) frequency   (2) period   (3) velocity   (4) wavelength   25 _____

26. An electromagnetic AM-band radio wave could have a wavelength of

(1) 0.005 m   (2) 5 m   (3) 500 m   (4) 5 000 000 m   26 _____

27. The accompanying diagram represents a transverse wave. The wavelength of the wave is equal to the distance between points

(1) A and G (3) C and E

(2) B and F (4) D and F 27 _____

28. When a light wave enters a new medium and is refracted, there must be a change in the light wave's

(1) color   (2) frequency   (3) period   (4) speed   28 _____

29. The speed of light in a piece of plastic is 2.00×10^8 meters per second. What is the absolute index of refraction of this plastic?

(1) 1.00   (2) 0.670   (3) 1.33   (4) 1.50   29 _____

30. Wave X travels eastward with frequency f and amplitude A. Wave Y, traveling in the same medium, interacts with wave X and produces a standing wave. Which statement about wave Y is correct?

(1) Wave Y must have a frequency of f, an amplitude of A, and be traveling eastward.

(2) Wave Y must have a frequency of $2f$, an amplitude of $3A$, and be traveling eastward.

(3) Wave Y must have a frequency of $3f$, an amplitude of $2A$, and be traveling westward.

(4) Wave Y must have a frequency of f, an amplitude of A, and be traveling westward.   30 _____

31. A car's horn is producing a sound wave having a constant frequency of 350 hertz. If the car moves toward a stationary observer at constant speed, the frequency of the car's horn detected by this observer may be

(1) 320 Hz   (2) 330 Hz   (3) 350 Hz   (4) 380 Hz   31 _____

32. The diagram below represents two pulses approaching each other from opposite directions in the same medium.

Which diagram best represents the medium after the pulses have passed through each other?

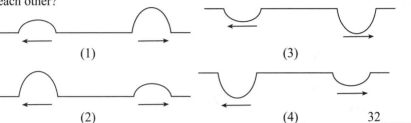

(1)   (3)

(2)   (4)   32 _____

33. A mercury atom in the ground state absorbs 20.00 electronvolts of energy and is ionized by losing an electron. How much kinetic energy does this electron have after the ionization?
(1) 6.40 eV   (2) 9.62 eV   (3) 10.38 eV   (4) 13.60 eV   33 _____

34. Which fundamental force is primarily responsible for the attraction between protons and electrons?
(1) strong   (2) weak   (3) gravitational   (4) electromagnetic   34 _____

35. The total conversion of 1.00 kilogram of the Sun's mass into energy yields
(1) 9.31×10^2 MeV   (3) 3.00×10^8 J
(2) 8.38×10^{19} MeV   (4) 9.00×10^{16} J   35 _____

Part B–1
Answer all questions in this part.

Directions (36–51): For *each* statement or question, write in the space provided the *number* of the word or expression that, of those given, best completes the statement or answers the question.

36. The accompanying graph represents the displacement of an object moving in a straight line as a function of time.

What was the total distance traveled by the object during the 10.0-second time interval?
(1) 0 m   (3) 16 m
(2) 8 m   (4) 24 m

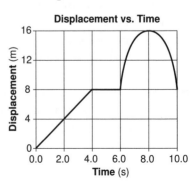

37. The mass of a paper clip is approximately
(1) 1×10^6 kg   (2) 1×10^3 kg   (3) 1×10^{-3} kg   (4) 1×10^{-6} kg   37_____

38. Which diagram best represents the gravitational forces, F_g, between a satellite, S, and Earth?

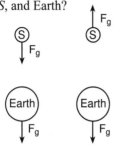 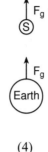

(1)   (2)   (3)   (4)   38_____

39. A block weighing 10.0 newtons is on a ramp inclined at 30.0° to the horizontal. A 3.0-newton force of friction, F_f, acts on the block as it is pulled up the ramp at constant velocity with force F, which is parallel to the ramp, as shown in the accompanying diagram. What is the magnitude of force F?

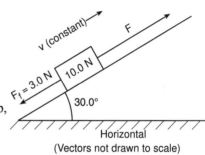

(Vectors not drawn to scale)

(1) 7.0 N   (3) 10. N
(2) 8.0 N   (4) 13 N   39_____

40. A 25-newton horizontal force northward and a 35-newton horizontal force southward act concurrently on a 15-kilogram object on a frictionless surface. What is the magnitude of the object's acceleration?
(1) 0.67 m/s²   (2) 1.7 m/s²   (3) 2.3 m/s²   (4) 4.0 m/s²   40_____

41. The accompanying diagram represents two concurrent forces. Which vector represents the force that will produce equilibrium with these two forces?

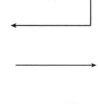

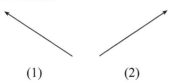

(1)   (2)   (3)   (3)   41_____

42. Which graph best represents the relationship between the magnitude of the centripetal acceleration and the speed of an object moving in a circle of constant radius?

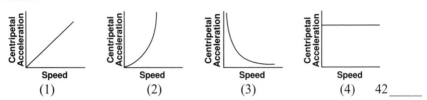

(1)   (2)   (3)   (4)   42 _____

43. The accompanying diagram represents two masses before and after they collide. Before the collision, mass m_A is moving to the right with speed v, and mass m_B is at rest. Upon collision, the two masses stick together.

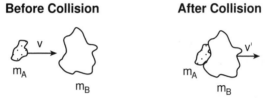

Which expression represents the speed, v', of the masses after the collision? [Assume no outside forces are acting on m_A or m_B.]

(1) $\dfrac{m_A + m_B v}{m_A}$ (2) $\dfrac{m_A + m_B}{m_A v}$ (3) $\dfrac{m_B v}{m_A + m_B}$ (4) $\dfrac{m_A v}{m_A + m_B}$ 43 _____

44. Which combination of fundamental units can be used to express energy?
(1) kg•m/s   (2) kg•m^2/s   (3) kg•m/s^2 (4) kg•m^2/s^2 44 _____

45. An object is thrown vertically upward. Which pair of graphs best represents the object's kinetic energy and gravitational potential energy as functions of its displacement while it rises?

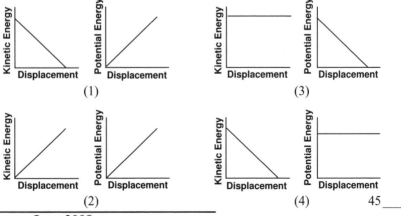

46. Charge flowing at the rate of 2.50×10^{16} elementary charges per second is equivalent to a current of

(1) 2.50×10^{13} A   (3) 4.00×10^{-3} A

(2) 6.25×10^{5} A   (4) 2.50×10^{-3} A   46 _____

47. An electric drill operating at 120. volts draws a current of 3.00 amperes. What is the total amount of electrical energy used by the drill during 1.00 minute of operation?

(1) 2.16×10^{4} J   (3) 3.60×10^{2} J

(2) 2.40×10^{3} J   (4) 4.00×10^{1} J   47 _____

48. The accompanying diagram represents a transverse wave traveling to the right through a medium. Point A represents a particle of the medium. In which direction will particle A move in the next instant of time?

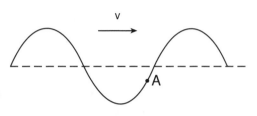

(1) up   (2) down   (3) left   (4) right   48 _____

49. Which graph best represents the relationship between photon energy and photon frequency?

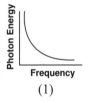 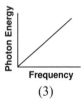 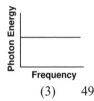

(1)   (2)   (3)   (3)   49 _____

Base your answers to questions 50 and 51 on the accompanying table, which shows data about various subatomic particles.

Subatomic Particle Table

Symbol	Name	Quark Content	Electric Charge	Mass (GeV/c²)
p	proton	uud	+1	0.938
p̄	antiproton	ūūd̄	−1	0.938
n	neutron	udd	0	0.940
λ	lambda	uds	0	1.116
Ω⁻	omega	sss	−1	1.672

50. Which particle listed on the table has the opposite charge of, and is more massive than, a proton?

(1) antiproton   (2) neutron   (3) lambda   (4) omega   50 _____

51. All the particles listed on the table are classified as

(1) mesons   (2) hadrons   (3) antimatter   (4) leptons   51 _____

Part B–2
Answer all questions in this part.
Directions (52–61): Record your answers in the spaces provided.

52. The accompanying graph represents the velocity of an object traveling in a straight line as a function of time. Determine the magnitude of the total displacement of the object at the end of the first 6.0 seconds. [1]

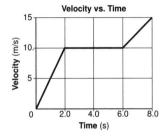

52 _____ m

Base your answers to questions 53 and 54 on the information below.

A 65-kilogram pole vaulter wishes to vault to a height of 5.5 meters.

53. Calculate the *minimum* amount of kinetic energy the vaulter needs to reach this height if air friction is neglected and all the vaulting energy is derived from kinetic energy. [Show all work, including the equation and substitution with units.] [2]

54. Calculate the speed the vaulter must attain to have the necessary kinetic energy. [Show all work, including the equation and substitution with units.] [2]

Base your answers to questions 55 through 57 on the information and accompanying vector diagram.

N
W ← → E
S

6.0 m

8.0 m

A dog walks 8.0 meters due north and then 6.0 meters due east.

55. Using a metric ruler and the vector determine the scale used in the diagram. [1]

1.0 cm = _____ m

56. On the diagram above, construct the resultant vector that represents the dog's total displacement. [1]

57. Determine the magnitude of the dog's total displacement. [1] _____ m

58. Two small identical metal spheres, A and B, on insulated stands, are each given a charge of $+2.0 \times 10^{-6}$ coulomb. The distance between the spheres is 2.0×10^{-1} meter. Calculate the magnitude of the electrostatic force that the charge on sphere A exerts on the charge on sphere B. [Show all work, including the equation and substitution with units.] [2]

Base your answers to questions 59 and 60 on the information and accompanying diagram.

A 10.0-meter length of copper wire is at 20°C. The radius of the wire is 1.0×10^{-3} meter.

Cross Section of Copper Wire

r = 1.0 × 10⁻³ m

59. Determine the cross-sectional area of the wire. [1] _____ m²

60. Calculate the resistance of the wire. [Show all work, including the equation and substitution with units.] [2]

61. The diagram below represents a transverse wave moving on a uniform rope with point A labeled as shown. On the diagram below, mark an **X** at the point on the wave that is 180° out of phase with point A. [1]

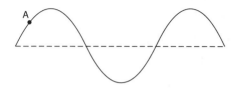

Part C
Answer all questions in this part.
Directions (62–76): **Record your answers in the spaces provided.**

Base your answers to questions 62 through 64 on the information below.

A kicked soccer ball has an initial velocity of 25 meters per second at an angle of 40.° above the horizontal, level ground. [Neglect friction.]

62. Calculate the magnitude of the vertical component of the ball's initial velocity. [Show all work, including the equation and substitution with units.] [2]

63. Calculate the maximum height the ball reaches above its initial position. [Show all work, including the equation and substitution with units.] [2]

64. On the diagram below, sketch the path of the ball's flight from its initial position at point *P* until it returns to level ground. [1]

P   Level ground

Base your answers to questions 65 through 67 on the information and accompanying diagram.

A 15-ohm resistor, R_1, and a 30.-ohm resistor, R_2, are to be connected in parallel between points *A* and *B* in a circuit containing a 90.-volt battery.

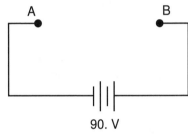

65. Complete the diagram above to show the two resistors connected in parallel between points *A* and *B*. [1]

66. Determine the potential difference across resistor R_1. [1]

_____ V

67. Calculate the current in resistor R_1. [Show all work, including the equation and substitution with units.] [2]

Base your answers to questions 68 through 71 on the information and data table below.

The spring in a dart launcher has a spring constant of 140 newtons per meter. The launcher has six power settings, 0 through 5, with each successive setting having a spring compression 0.020 meter beyond the previous setting. During testing, the launcher is aligned to the vertical, the spring is compressed, and a dart is fired upward. The maximum vertical displacement of the dart in each test trial is measured. The results of the testing are shown in the accompanying table.

Data Table

Power Setting	Spring Compression (m)	Dart's Maximum Vertical Displacement (m)
0	0.000	0.00
1	0.020	0.29
2	0.040	1.14
3	0.060	2.57
4	0.080	4.57
5	0.100	7.10

Directions (68–69): Using the information in the data table, construct a graph on the accompanying grid, following the directions below.

68. Plot the data points for the dart's maximum vertical displacement versus spring compression. [1]

69. Draw the line or curve of best fit. [1]

70. Using information from your graph, calculate the energy provided by the compressed spring that causes the dart to achieve a maximum vertical displacement of 3.50 meters. [Show all work, including the equation and substitution with units.] [2]

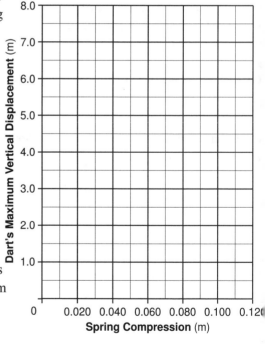

Dart's Maximum Vertical Displacement vs. Spring Compression

71. Determine the magnitude of the force, in
newtons, needed to compress the spring 0.040 meter. [1] _____ N

Base your answers to questions 72 through 74 on the information and diagram below.

A ray of monochromatic light having a frequency of 5.09×10^{14} hertz is incident on an interface of air and corn oil at an angle of 35° as shown. The ray is transmitted through parallel layers of corn oil and glycerol and is then reflected from the surface of a plane mirror, located below and parallel to the glycerol layer. The ray then emerges from the corn oil back into the air at point P.

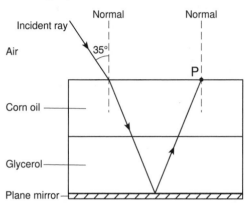

72. Calculate the angle of refraction of the light ray as it enters the corn oil from air. [Show all work, including the equation and the substitution with units.] [2]

73. Explain why the ray does not bend at the corn oil-glycerol interface. [1]

_____

_____

_____

74. On the diagram above, use a protractor and straightedge to construct the refracted ray representing the light emerging at point P into air. [1]

Base your answers to questions 75 and 76 on the information and data table below.

In the first nuclear reaction using a particle accelerator, accelerated protons bombarded lithium atoms, producing alpha particles and energy. The energy resulted from the conversion of mass into energy. The reaction can be written as shown below.

$$^1_1H + {}^7_3Li \rightarrow {}^4_2He + {}^4_2He + energy$$

Data Table

Particle	Symbol	Mass (u)
proton	^1_1H	1.007 83
lithium atom	^7_3Li	7.016 00
alpha particle	^4_2He	4.002 60

75. Determine the difference between the total mass of a proton plus a lithium atom, $^1_1H + {}^7_3Li$ and the total mass of two alpha particles, $^4_2He + {}^4_2He$, in universal mass units. [1] _____ u

76. Determine the energy in megaelectronvolts produced in the reaction of a proton with a lithium atom. [1] _____ MeV

June 2009
Part A
Answer all questions in this part.

Directions (1–35): For *each* statement or question, write in the space provided the *number* of the word or expression that, of those given, best completes the statement or answers the question.

1. On a highway, a car is driven 80. kilometers during the first 1.00 hour of travel, 50. kilometers during the next 0.50 hour, and 40. kilometers in the final 0.50 hour. What is the car's average speed for the entire trip?
(1) 45 km/h   (2) 60. km/h   (3) 85 km/h   (4) 170 km/h   1_____

2. The accompanying vector diagram represents the horizontal component, F_H, and the vertical component, F_V, of a 24-newton force acting at 35° above the horizontal. What are the magnitudes of the horizontal and vertical components?
(1) $F_H = 3.5$ N and $F_V = 4.9$ N
(2) $F_H = 4.9$ N and $F_V = 3.5$ N
(3) $F_H = 14$ N and $F_V = 20.$ N
(4) $F_H = 20.$ N and $F_V = 14$ N   2_____

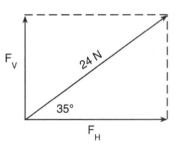

3. Which quantity is a vector?
(1) impulse   (2) power   (3) speed   (4) time   3_____

4. A high-speed train in Japan travels a distance of 300. kilometers in 3.60×10^3 seconds. What is the average speed of this train?
(1) 1.20×10^{-2} m/s   (3) 12.0 m/s
(2) 8.33×10^{-2} m/s   (4) 83.3 m/s   4_____

5. A 25-newton weight falls freely from rest from the roof of a building. What is the total distance the weight falls in the first 1.0 second?
(1) 19.6 m   (2) 9.8 m   (3) 4.9 m   (4) 2.5 m   5_____

6. A golf ball is given an initial speed of 20. meters per second and returns to level ground. Which launch angle above level ground results in the ball traveling the greatest horizontal distance? [Neglect friction.]
(1) 60.°   (2) 45°   (3) 30.°   (4) 15°   6_____

Base your answers to questions 7 and 8 on the information below.

A go-cart travels around a flat, horizontal, circular track with a radius of 25 meters. The mass of the go-cart with the rider is 200. kilograms. The magnitude of the maximum centripetal force exerted by the track on the go-cart is 1200. newtons.

7. What is the maximum speed the 200.-kilogram go-cart can travel without sliding off the track?
(1) 8.0 m/s   (2) 12 m/s   (3) 150 m/s   (4) 170 m/s   7_____

8. Which change would increase the maximum speed at which the go-cart could travel without sliding off this track?
(1) Decrease the coefficient of friction between the go-cart and the track.
(2) Decrease the radius of the track.
(3) Increase the radius of the track.
(4) Increase the mass of the go-cart.   8_____

9. A 0.50-kilogram cart is rolling at a speed of 0.40 meter per second. If the speed of the cart is doubled, the inertia of the cart is
(1) halved   (2) doubled   (3) quadrupled   (4) unchanged   9_____

10. Two forces, F_1 and F_2, are applied to a block on a frictionless, horizontal surface as shown below.

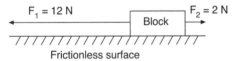

If the magnitude of the block's acceleration is 2.0 meters per second², what is the mass of the block?
(1) 1 kg   (2) 5 kg   (3) 6 kg   (4) 7 kg   10_____

11. Which body is in equilibrium?
(1) a satellite orbiting Earth in a circular orbit
(2) a ball falling freely toward the surface of Earth
(3) a car moving with a constant speed along a straight, level road
(4) a projectile at the highest point in its trajectory   11_____

12. What is the weight of a 2.00-kilogram object on the surface of Earth?
(1) 4.91 N   (2) 2.00 N   (3) 9.81 N   (4) 19.6 N   12_____

13. A 70.-kilogram cyclist develops 210 watts of power while pedaling at a constant velocity of 7.0 meters per second east. What average force is exerted eastward on the bicycle to maintain this constant speed?
(1) 490 N   (2) 30. N   (3) 3.0 N   (4) 0 N   13____

14. The gravitational potential energy, with respect to Earth, that is possessed by an object is dependent on the object's
(1) acceleration   (2) momentum   (3) position   (4) speed   14____

Note that question 15 has only three choices.

15. As a ball falls freely toward the ground, its total mechanical energy
(1) decreases   (2) increases   (3) remains the same   15____

16. A spring with a spring constant of 4.0 newtons per meter is compressed by a force of 1.2 newtons. What is the total elastic potential energy stored in this compressed spring?
(1) 0.18 J   (2) 0.36 J   (3) 0.60 J   (4) 4.8 J   16____

17. A distance of 1.0 meter separates the centers of two small charged spheres. The spheres exert gravitational force F_g and electrostatic force F_e on each other. If the distance between the spheres' centers is increased to 3.0 meters, the gravitational force and electrostatic force, respectively, may be represented as
(1) $\dfrac{F_g}{9}$ and $\dfrac{F_e}{9}$ (3) $3F_g$ and $3F_e$

(2) $\dfrac{F_g}{3}$ and $\dfrac{F_e}{3}$ (4) $9F_g$ and $9F_e$ 17____

18. The electrical resistance of a metallic conductor is inversely proportional to its
(1) temperature   (3) cross-sectional area
(2) length   (4) resistivity   18____

19. In a simple electric circuit, a 24-ohm resistor is connected across a 6.0-volt battery. What is the current in the circuit?
(1) 1.0 A   (2) 0.25 A   (3) 140 A   (4) 4.0 A   19____

20. An operating 100.-watt lamp is connected to a 120-volt outlet. What is the total electrical energy used by the lamp in 60. seconds?
(1) 0.60 J   (2) 1.7 J   (3) 6.0×10^3 J   (4) 7.2×10^3 J   20____

21. A beam of electrons is directed into the electric field between two oppositely charged parallel plates, as shown in the diagram below.

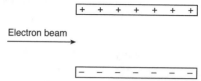

The electrostatic force exerted on the electrons by the electric field is directed
(1) into the page
(2) out of the page
(3) toward the bottom of the page
(4) toward the top of the page   21_____

22. When two ring magnets are placed on a pencil, magnet *A* remains suspended above magnet *B*, as shown to the right. Which statement describes the gravitational force and the magnetic force acting on magnet *A* due to magnet *B*?

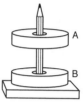

(1) The gravitational force is attractive and the magnetic force is repulsive.
(2) The gravitational force is repulsive and the magnetic force is attractive.
(3) Both the gravitational force and the magnetic force are attractive.
(4) Both the gravitational force and the magnetic force are repulsive.   22_____

23. Which color of light has a wavelength of 5.0×10^{-7} meter in air?
(1) blue   (2) green   (3) orange   (4) violet   23_____

24. Which type of wave requires a material medium through which to travel?
(1) sound   (2) radio   (3) television   (4) x ray   24_____

25. A periodic wave is produced by a vibrating tuning fork. The amplitude of the wave would be greater if the tuning fork were
(1) struck more softly   (3) replaced by a lower frequency tuning fork
(2) struck harder   (4) replaced by a higher frequency tuning fork   25_____

26. The sound wave produced by a trumpet has a frequency of 440 hertz. What is the distance between successive compressions in this sound wave as it travels through air at STP?
(1) 1.5×10^{-6} m   (2) 0.75 m   (3) 1.3 m   (4) 6.8×10^{5} m   26_____

27. The accompanying diagram represents a light ray striking the boundary between air and glass. What would be the angle between this light ray and its reflected ray?

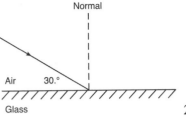

(1) 30.° (3) 120.°

(2) 60.° (4) 150.°

27_____

28. In which way does blue light change as it travels from diamond into crown glass?

(1) Its frequency decreases. (3) Its speed decreases.

(2) Its frequency increases. (4) Its speed increases. 28_____

29. The diagram below shows two pulses approaching each other in a uniform medium.

Which diagram best represents the superposition of the two pulses?

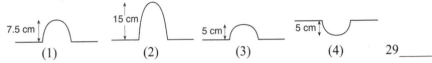

29_____

30. Sound waves strike a glass and cause it to shatter. This phenomenon illustrates

(1) resonance (2) refraction (3) reflection (4) diffraction 30_____

31. An alpha particle consists of two protons and two neutrons. What is the charge of an alpha particle?

(1) 1.25×10^{19} C (2) 2.00 C (3) 6.40×10^{-19} C (4) 3.20×10^{-19} C 31_____

32. An electron in the c level of a mercury atom returns to the ground state. Which photon energy could *not* be emitted by the atom during this process?

(1) 0.22 eV (2) 4.64 eV (3) 4.86 eV (4) 5.43 eV 32_____

33. Which phenomenon provides evidence that light has a wave nature?

(1) emission of light from an energy-level transition in a hydrogen atom

(2) diffraction of light passing through a narrow opening

(3) absorption of light by a black sheet of paper

(4) reflection of light from a mirror 33_____

34. When Earth and the Moon are separated by a distance of 3.84×10^8 meters, the magnitude of the gravitational force of attraction between them is 2.0×10^{20} newtons. What would be the magnitude of this gravitational force of attraction if Earth and the Moon were separated by a distance of 1.92×10^8 meters?

(1) 5.0×10^{19} N

(3) 4.0×10^{20} N

(2) 2.0×10^{20} N

(4) 8.0×10^{20} N

34_____

35. The particles in a nucleus are held together primarily by the

(1) strong force

(3) electrostatic force

(2) gravitational force

(4) magnetic force

35_____

Part B–1

Answer all questions in this part.

Directions **(36–47): For *each* statement or question, write in the space provided the *number* of the word or expression that, of those given, best completes the statement or answers the question.**

36. The work done in lifting an apple one meter near Earth's surface is approximately

(1) 1 J (2) 0.01 J (3) 100 J (4) 1000 J

36_____

Base your answers to questions 37 and 38 on the accompanying graph, which represents the motion of a car during a 6.0-second time interval.

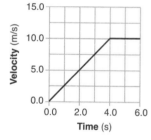

37. What is the acceleration of the car at $t = 5.0$ seconds?

(1) 0.0 m/s² (3) 2.5 m/s²

(2) 2.0 m/s² (4) 10. m/s²

37_____

38. What is the total distance traveled by the car during this 6.0-second interval?

(1) 10. m (2) 20. m (3) 40. m (4) 60. m

38_____

39. A person weighing 785 newtons on the surface of Earth would weigh 298 newtons on the surface of Mars. What is the magnitude of the gravitational field strength on the surface of Mars?

(1) 2.63 N/kg (2) 3.72 N/kg (3) 6.09 N/kg (4) 9.81 N/kg

39_____

40. A motorcycle being driven on a dirt path hits a rock. Its 60.-kilogram cyclist is projected over the handlebars at 20. meters per second into a haystack. If the cyclist is brought to rest in 0.50 second, the magnitude of the average force exerted on the cyclist by the haystack is
(1) 6.0×10^1 N   (2) 5.9×10^2 N   (3) 1.2×10^3 N   (4) 2.4×10^3 N   40____

Base your answers to questions 41 and 42 on the information below.

A boy pushes his wagon at constant speed along a level sidewalk. The graph below represents the relationship between the horizontal force exerted by the boy and the distance the wagon moves.

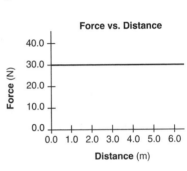

41. What is the total work done by the boy in pushing the wagon 4.0 meters?
(1) 5.0 J   (3) 120 J
(2) 7.5 J   (4) 180 J   41____

42. As the boy pushes the wagon, what happens to the wagon's energy?
(1) Gravitational potential energy increases.
(2) Gravitational potential energy decreases.
(3) Internal energy increases.
(4) Internal energy decreases.
                                            42____

43. Which is an SI unit for work done on an object?
(1) $\dfrac{kg \cdot m^2}{s^2}$ (2) $\dfrac{kg \cdot m^2}{s}$ (3) $\dfrac{kg \cdot m}{s}$ (4) $\dfrac{kg \cdot m}{s^2}$ 43____

44. The momentum of a photon, p, is given by the equation $p = \dfrac{h}{\lambda}$ where h is Planck's constant and λ is the photon's wavelength. Which equation correctly represents the energy of a photon in terms of its momentum?

(1) $E_{photon} = phc$ (2) $E_{photon} = \dfrac{hp}{c}$ (3) $E_{photon} = \dfrac{p}{c}$ (4) $E_{photon} = pc$ 44____

45. A constant potential difference is applied across a variable resistor held at constant temperature. Which graph best represents the relationship between the resistance of the variable resistor and the current through it?

(1)

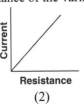
(2)

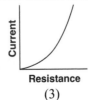
(3)

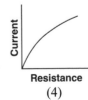
(4)   45____

46. A 3.0-ohm resistor and a 6.0-ohm resistor are connected in series in an operating electric circuit. If the current through the 3.0-ohm resistor is 4.0 amperes, what is the potential difference across the 6.0-ohm resistor?
(1) 8.0 V   (2) 2.0 V   (3) 12 V   (4) 24 V   46____

47. Which combination of resistors has the *smallest* equivalent resistance?

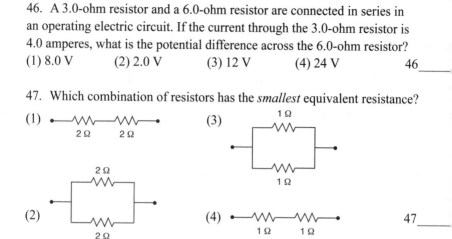

Part B–2
Answer all questions in this part.
Directions **(48–59): Record your answers in the spaces provided.**

48. A cart travels 4.00 meters east and then 4.00 meters north. Determine the magnitude of the cart's resultant displacement. [1] _____ m

49. A 70-kilogram hockey player skating east on an ice rink is hit by a 0.1-kilogram hockey puck moving toward the west. The puck exerts a 50-newton force toward the west on the player. Determine the magnitude of the force that the player exerts on the puck during this collision. [1] _____ N

50. On a snow-covered road, a car with a mass of 1.1×10^3 kilograms collides head-on with a van having a mass of 2.5×10^3 kilograms traveling at 8.0 meters per second. As a result of the collision, the vehicles lock together and immediately come to rest. Calculate the speed of the car immediately before the collision. [Neglect friction.] [Show all work, including the equation and substitution with units.] [2]

51. A baby and stroller have a total mass of 20. kilograms. A force of 36 newtons keeps the stroller moving in a circular path with a radius of 5.0 meters. Calculate the speed at which the stroller moves around the curve. [Show all work, including the equation and substitution with units.] [2]

52. A 10.-newton force compresses a spring 0.25 meter from its equilibrium position. Calculate the spring constant of this spring. [Show all work, including the equation and substitution with units.] [2]

53. Two oppositely charged parallel metal plates, 1.00 centimeter apart, exert a force with a magnitude of 3.60×10^{-15} newton on an electron placed between the plates. Calculate the magnitude of the electric field strength between the plates. [Show all work, including the equation and substitution with units.] [2]

54. On the diagram below, sketch *at least four* electric field lines with arrowheads that represent the electric field around a negatively charged conducting sphere. [1]

$\ominus$

55. In the space below, draw a diagram of an operating circuit that includes:
   • a battery as a source of potential difference
   • *two* resistors in parallel with each other
   • an ammeter that reads the total current in the circuit [2]

56. Calculate the resistance of a 900.-watt toaster operating at 120 volts. [Show all work, including the equation and substitution with units.] [2]

57. A student and a physics teacher hold opposite ends of a horizontal spring stretched from west to east along a tabletop. Identify the directions in which the student should vibrate the end of the spring to produce transverse periodic waves. [1]

_____

_____

_____

Base your answers to questions 58 and 59 on the information and diagram below.

The vertical lines in the diagram represent compressions in a sound wave of constant frequency propagating to the right from a speaker toward an observer at point *A*.

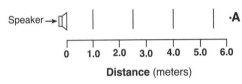

58. Determine the wavelength of this sound wave. [1]

58 _____ **m**

59. The speaker is then moved at constant speed toward the observer at *A*. Compare the wavelength of the sound wave received by the observer while the speaker is moving to the wavelength observed when the speaker was at rest. [1]

_____

_____

_____

Part C
Answer all questions in this part.
Directions (60–72): Record your answers in the spaces provided.

Base your answers to questions 60 through 62 on the information below.

The path of a stunt car driven horizontally off a cliff is represented in the accompanying diagram. After leaving the cliff, the car falls freely to point A in 0.50 second and to point B in 1.00 second.

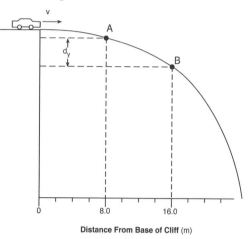

Distance From Base of Cliff (m)

60. Determine the magnitude of the horizontal component of the velocity of the car at point B. [Neglect friction.] [1]

60 _____ **m/s**

61. Determine the magnitude of the vertical velocity of the car at point A. [1]

61 _____ **m/s**

62. Calculate the magnitude of the vertical displacement, d_y, of the car from point A to point B. [Neglect friction.] [Show all work, including the equation and substitution with units.] [2]

Base your answers to questions 63 through 65 on the information below.

A roller coaster car has a mass of 290. kilograms. Starting from rest, the car acquires 3.13×10^5 joules of kinetic energy as it descends to the bottom of a hill in 5.3 seconds.

63. Calculate the height of the hill. [Neglect friction.] [Show all work, including the equation and substitution with units.] [2]

64. Calculate the speed of the roller coaster car at the bottom of the hill. [Show all work, including the equation and substitution with units.] [2]

65. Calculate the magnitude of the average acceleration of the roller coaster car as it descends to the bottom of the hill. [Show all work, including the equation and substitution with units.] [2]

Base your answers to questions 66 and 67 on the information below.

One end of a rope is attached to a variable speed drill and the other end is attached to a 5.0-kilogram mass. The rope is draped over a hook on a wall opposite the drill. When the drill rotates at a frequency of 20.0 Hz, standing waves of the same frequency are set up in the rope. The diagram below shows such a wave pattern.

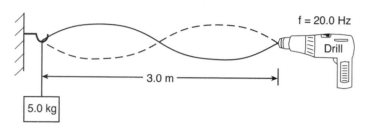

66. Determine the wavelength of the waves producing the standing wave pattern. [1]

66 _____ m

67. Calculate the speed of the wave in the rope. [Show all work, including the equation and substitution with units.] [2]

Base your answers to questions 68 and 69 on the information below.

A ray of monochromatic light ($f = 5.09 \times 10^{14}$ Hz) passes from air into Lucite at an angle of incidence of 30.°.

68. Calculate the angle of refraction in the Lucite. [Show all work, including the equation and substitution with units.] [2]

69. Using a protractor and straightedge, on the diagram below, draw the refracted ray in the Lucite. [1]

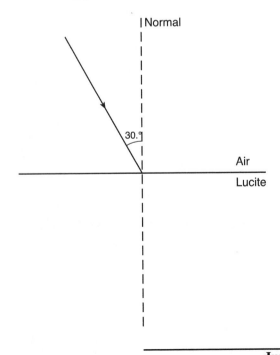

Base your answers to questions 70 through 72 on the information below.

A photon with a frequency of 5.48×10^{14} hertz is emitted when an electron in a mercury atom falls to a lower energy level.

70. Identify the color of light associated with this photon. [1]

70 _____

71. Calculate the energy of this photon in joules. [Show all work, including the equation and substitution with units.] [2]

72. Determine the energy of this photon in electronvolts. [1]

72 _____ eV

Directions (1–35): For *each* statement or question, write in the space provide the *number* of the word or expression that, of those given, best completes the statement or answers the question.

1. A baseball player runs 27.4 meters from the batter's box to first base, overruns first base by 3.0 meters, and then returns to first base. Compared to the total distance traveled by the player, the magnitude of the player's total displacement from the batter's box is
(1) 3.0 m shorter   (3) 3.0 m longer
(2) 6.0 m shorter   (4) 6.0 m longer   1 _____

2. A motorboat, which has a speed of 5.0 meters per second in still water, is headed east as it crosses a river flowing south at 3.3 meters per second. What is the magnitude of the boat's resultant velocity with respect to the starting point?
(1) 3.3 m/s   (2) 5.0 m/s   (3) 6.0 m/s   (4) 8.3 m/s   2 _____

3. A car traveling on a straight road at 15.0 meters per second accelerates uniformly to a speed of 21.0 meters per second in 12.0 seconds. The total distance traveled by the car in this 12.0-second time interval is
(1) 36.0 m   (2) 180. m   (3) 216 m   (4) 252 m   3 _____

4. A 0.149-kilogram baseball, initially moving at 15 meters per second, is brought to rest in 0.040 second by a baseball glove on a catcher's hand. The magnitude of the average force exerted on the ball by the glove is
(1) 2.2 N   (2) 2.9 N   (3) 17 N   (4) 56 N   4 _____

5. As shown in the accompanying diagram, a student standing on the roof of a 50.0-meter-high building kicks a stone at a horizontal speed of 4.00 meters per second.
How much time is required for the stone to reach the level ground below?
[Neglect friction.]
(1) 3.19 s   (3) 10.2 s
(2) 5.10 s   (4) 12.5 s

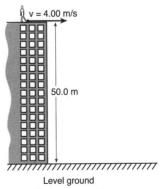

v = 4.00 m/s

50.0 m

Level ground
(Not drawn to scale)

5 _____

6. Which body is in equilibrium?
(1) a satellite moving around Earth in a circular orbit
(2) a cart rolling down a frictionless incline
(3) an apple falling freely toward the surface of Earth
(4) a block sliding at constant velocity across a tabletop   6 _____

7. On the surface of Earth, a spacecraft has a mass of 2.00×10^4 kilograms. What is the mass of the spacecraft at a distance of one Earth radius above Earth's surface?
(1) 5.00×10^3 kg   (3) 4.90×10^4 kg
(2) 2.00×10^4 kg   (4) 1.96×10^5 kg   7 _____

8. A student pulls a 60.-newton sled with a force having a magnitude of 20. newtons. What is the magnitude of the force that the sled exerts on the student?
(1) 20. N   (2) 40. N   (3) 60. N   (4) 80. N   8 _____

9. The accompanying data table lists the mass and speed of four different objects. Which object has the greatest inertia?
(1) A   (3) C
(2) B   (4) D

Data Table

Object	Mass (kg)	Speed (m/s)
A	4.0	6.0
B	6.0	5.0
C	8.0	3.0
D	16.0	1.5

9 _____

10. The accompanying diagram shows a horizontal 12-newton force being applied to two blocks, A and B, initially at rest on a horizontal, frictionless surface. Block A has a mass of 1.0 kilogram and block B has a mass of 2.0 kilograms.

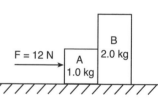

The magnitude of the acceleration of block B is
(1) 6.0 m/s²   (2) 2.0 m/s²   (3) 3.0 m/s²   (4) 4.0 m/s²   10 _____

11. A ball is thrown vertically upward with an initial velocity of 29.4 meters per second. What is the maximum height reached by the ball? [Neglect friction.]
(1) 14.7 m   (2) 29.4 m   (3) 44.1 m   (4) 88.1 m   11 _____

12. The accompanying diagram represents a mass, m, being swung clockwise at constant speed in a horizontal circle. At the instant shown, the centripetal force acting on mass m is directed toward point

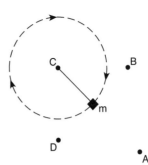

(1) A (3) C
(2) B (4) D

12_____

13. A 3.1-kilogram gun initially at rest is free to move. When a 0.015-kilogram bullet leaves the gun with a speed of 500. meters per second, what is the speed of the gun?

(1) 0.0 m/s   (2) 2.4 m/s   (3) 7.5 m/s   (4) 500. m/s   13 _____

14. Four projectiles, A, B, C, and D, were launched from, and returned to, level ground. The data table below shows the initial horizontal speed, initial vertical speed, and time of flight for each projectile.

Data Table

Projectile	Initial Horizontal Speed (m/s)	Initial Vertical Speed (m/s)	Time of Flight (s)
A	40.0	29.4	6.00
B	60.0	19.6	4.00
C	50.0	24.5	5.00
D	80.0	19.6	4.00

Which projectile traveled the greatest horizontal distance? [Neglect friction.]

(1) A (2) B (3) C (4) D 14 _____

15. The diagram below represents a 155-newton box on a ramp. Applied force F causes the box to slide from point A to point B.

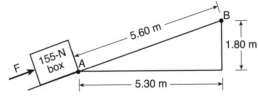

What is the total amount of gravitational potential energy gained by the box?

(1) 28.4 J   (2) 279 J   (3) 868 J   (4) 2740 J   15 _____

16. A wound spring provides the energy to propel a toy car across a level floor. At time t_i, the car is moving at speed v_i across the floor and the spring is unwinding, as shown below. At time t_f, the spring has fully unwound and the car has coasted to a stop.

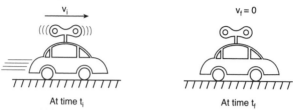

At time t_i At time t_f

Which statement best describes the transformation of energy that occurs between times t_i and t_f?
(1) Gravitational potential energy at t_i is converted to internal energy at t_f.
(2) Elastic potential energy at t_i is converted to kinetic energy at t_f.
(3) Both elastic potential energy and kinetic energy at ti are converted to internal energy at t_f.
(4) Both kinetic energy and internal energy at t_i are converted to elastic potential energy at t_f.

16 _____

17. A 75-kilogram bicyclist coasts down a hill at a constant speed of 12 meters per second. What is the kinetic energy of the bicyclist?
(1) 4.5×10^2 J (2) 9.0×10^2 J (3) 5.4×10^3 J (4) 1.1×10^4 J

17 _____

18. An electric heater operating at 120. volts draws 8.00 amperes of current through its 15.0 ohms of resistance. The total amount of heat energy produced by the heater in 60.0 seconds is
(1) 7.20×10^3 J (2) 5.76×10^4 J (3) 8.64×10^4 J (4) 6.91×10^6 J

18 _____

19. Magnetic fields are produced by particles that are
(1) moving and charged   (3) stationary and charged
(2) moving and neutral   (4) stationary and neutral

19 _____

20. A charge of 30. coulombs passes through a 24-ohm resistor in 6.0 seconds. What is the current through the resistor?
(1) 1.3 A   (2) 5.0 A   (3) 7.5 A   (4) 4.0 A

20 _____

21. What is the magnitude of the electrostatic force between two electrons separated by a distance of 1.00×10^{-8} meter?
(1) 2.56×10^{-22} N   (3) 2.30×10^{-12} N
(2) 2.30×10^{-20} N   (4) 1.44×10^{-1} N

21 _____

22. The accompanying diagram represents the electric field surrounding two charged spheres, A and B. What is the sign of the charge of each sphere?
(1) Sphere A is positive and sphere B is negative.
(2) Sphere A is negative and sphere B is positive.
(3) Both spheres are positive.
(4) Both spheres are negative.

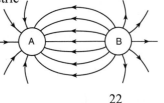

22 _____

23. Which circuit has the *smallest* equivalent resistance?

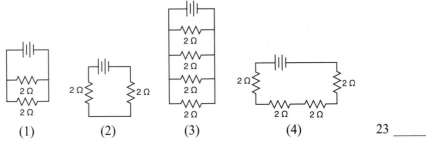

(1)   (2)   (3)   (4)   23 _____

Base your answers to questions 24 through 26 on the information and diagram below.

A longitudinal wave moves to the right through a uniform medium, as shown below. Points A, B, C, D, and E represent the positions of particles of the medium.

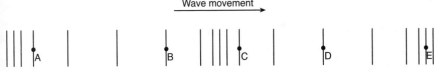

Wave movement

24. Which diagram best represents the motion of the particle at position C as the wave moves to the right?

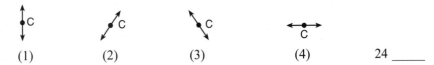

(1)   (2)   (3)   (4)   24 _____

25. The wavelength of this wave is equal to the distance between points
(1) A and B   (2) A and C   (3) B and C   (4) B and E   25 _____

26. The energy of this wave is related to its
(1) amplitude   (2) period   (3) speed   (4) wavelength   26 _____

27. A ray of monochromatic yellow light ($f = 5.09 \times 10^{14}$ Hz) passes from water through flint glass and into medium X, as shown to the right. The absolute index of refraction of medium X is

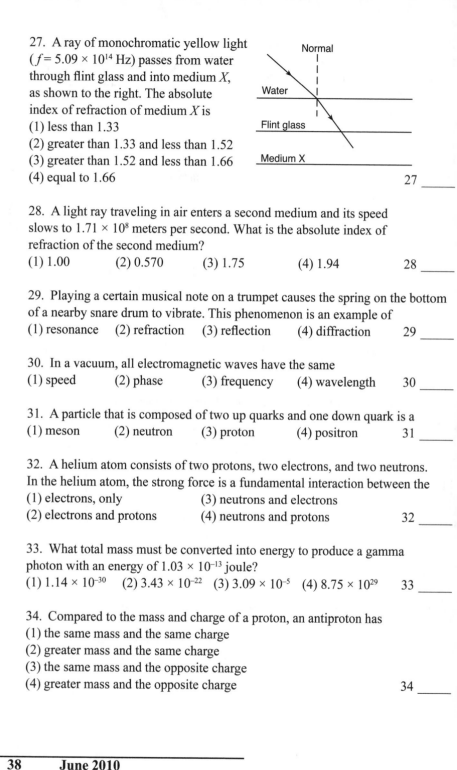

Normal

Water

Flint glass

Medium X

(1) less than 1.33
(2) greater than 1.33 and less than 1.52
(3) greater than 1.52 and less than 1.66
(4) equal to 1.66

27 _____

28. A light ray traveling in air enters a second medium and its speed slows to 1.71×10^8 meters per second. What is the absolute index of refraction of the second medium?

(1) 1.00   (2) 0.570   (3) 1.75   (4) 1.94   28 _____

29. Playing a certain musical note on a trumpet causes the spring on the bottom of a nearby snare drum to vibrate. This phenomenon is an example of

(1) resonance   (2) refraction   (3) reflection   (4) diffraction   29 _____

30. In a vacuum, all electromagnetic waves have the same

(1) speed   (2) phase   (3) frequency   (4) wavelength   30 _____

31. A particle that is composed of two up quarks and one down quark is a

(1) meson   (2) neutron   (3) proton   (4) positron   31 _____

32. A helium atom consists of two protons, two electrons, and two neutrons. In the helium atom, the strong force is a fundamental interaction between the

(1) electrons, only   (3) neutrons and electrons
(2) electrons and protons   (4) neutrons and protons   32 _____

33. What total mass must be converted into energy to produce a gamma photon with an energy of 1.03×10^{-13} joule?

(1) 1.14×10^{-30} (2) 3.43×10^{-22} (3) 3.09×10^{-5} (4) 8.75×10^{29} 33 _____

34. Compared to the mass and charge of a proton, an antiproton has

(1) the same mass and the same charge
(2) greater mass and the same charge
(3) the same mass and the opposite charge
(4) greater mass and the opposite charge

34 _____

Note that question 35 has only three choices.

35. As viewed from Earth, the light from a star has lower frequencies than the light emitted by the star because the star is
(1) moving toward Earth   (2) moving away from Earth   (3) stationary   35 _____

Part B–1
Answer all questions in this part.
Directions (36–50): For *each* statement or question, write in the space provided the *number* of the word or expression that, of those given, best completes the statement or answers the question.

36. The total work done in lifting a typical high school physics textbook a vertical distance of 0.10 meter is approximately
(1) 0.15 J   (2) 1.5 J   (3) 15 J   (4) 150 J   36 _____

37. Which electrical unit is equivalent to one joule?
(1) volt per meter   (3) volt per coulomb
(2) ampere•volt   (4) coulomb•volt   37 _____

38. A small electric motor is used to lift a 0.50-kilogram mass at constant speed. If the mass is lifted a vertical distance of 1.5 meters in 5.0 seconds, the average power developed by the motor is
(1) 0.15 W   (2) 1.5 W   (3) 3.8 W   (4) 7.5 W   38 _____

39. A ball is dropped from the top of a cliff. Which graph best represents the relationship between the ball's total energy and elapsed time as the ball falls to the ground? [Neglect friction.]

(1)   (2)   (3)   (4)   39 _____

40. A child, starting from rest at the top of a playground slide, reaches a speed of 7.0 meters per second at the bottom of the slide. What is the vertical height of the slide? [Neglect friction.]
(1) 0.71 m   (2) 1.4 m   (3) 2.5 m   (4) 3.5 m   40 _____

41. The accompanying graph represents the relationship between the current in a metallic conductor and the potential difference across the conductor at constant temperature. The resistance of the conductor is

(1) 1.0 Ω   (3) 0.50 Ω

(2) 2.0 Ω   (4) 4.0 Ω

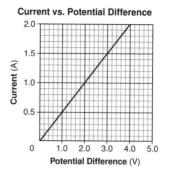

Current vs. Potential Difference

41 _____

42. A student throws a baseball vertically upward and then catches it. If vertically upward is considered to be the positive direction, which graph best represents the relationship between velocity and time for the baseball? [Neglect friction.]

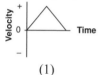

    (1)   (2)   (3)   (4)

42 _____

43. A 5.0-kilogram sphere, starting from rest, falls freely 22 meters in 3.0 seconds near the surface of a planet. Compared to the acceleration due to gravity near Earth's surface, the acceleration due to gravity near the surface of the planet is approximately

(1) the same   (3) one-half as great

(2) twice as great   (4) four times as great

43 _____

44. A 15.0-kilogram mass is moving at 7.50 meters per second on a horizontal, frictionless surface. What is the total work that must be done on the mass to increase its speed to 11.5 meters per second?

(1) 120. J   (2) 422 J   (3) 570. J   (4) 992 J

44 _____

45. The accompanying circuit diagram represents four resistors connected to a 12-volt source. What is the total current in the circuit?

(1) 0.50 A   (3) 8.6 A

(2) 2.0 A   (4) 24 A

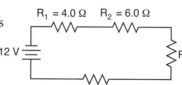

45 _____

46. Which graph best represents the relationship between the power expended by a resistor that obeys Ohm's Law and the potential difference applied to the resistor?

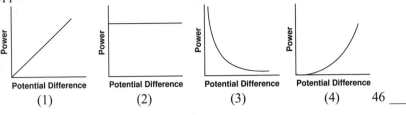

| (1) | (2) | (3) | (4) | 46 _____ |

47. The distance between an electron and a proton is varied. Which pair of graphs best represents the relationship between gravitational force, F_g, and distance, r, and the relationship between electrostatic force, F_e, and distance, r, for these particles?

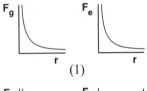

(1)

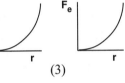

(3)

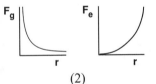

(2)

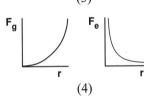

(4)   47 _____

48. The accompanying diagram represents a periodic wave traveling through a uniform medium. If the frequency of the wave is 2.0 hertz, the speed of the wave is

(1) 6.0 m/s   (2) 2.0 m/s   (3) 8.0 m/s   (4) 4.0 m/s   48 _____

49. The accompanying diagram represents a light ray reflecting from a plane mirror. The angle of reflection for the light ray is

(1) 25°   (3) 50.°
(2) 35°   (4) 65°

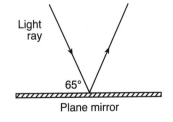

49 _____

50. The accompanying diagram shows a standing wave in a string clamped at each end. What is the total number of nodes and antinodes in the standing wave?
(1) 3 nodes and 2 antinodes
(2) 2 nodes and 3 antinodes
(3) 5 nodes and 4 antinodes
(4) 4 nodes and 5 antinodes

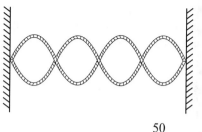

50 _____

Part B–2
Answer all questions in this part.
Directions **(51–60): Record your answers in the spaces provided.**

Base your answers to questions 51 through 53 on the information and graph below.

A machine fired several projectiles at the same angle, θ, above the horizontal. Each projectile was fired with a different initial velocity, v_i. The accompanying graph represents the relationship between the magnitude of the initial vertical velocity, v_{iy}, and the magnitude of the corresponding initial velocity, v_i, of these projectiles.

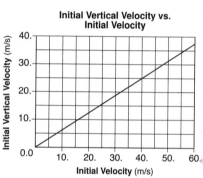

51. Determine the magnitude of the initial vertical velocity of the projectile, v_{iy}, when the magnitude of its initial velocity, v_i, was 40. meters per second. [1]   _____m/s

52. Determine the angle, θ, above the horizontal at which the projectiles were fired. [1]   _____°

53. Calculate the magnitude of the initial horizontal velocity of the projectile, v_{ix}, when the magnitude of its initial velocity, v_i, was 40. meters per second. [Show all work, including the equation and substitution with units.] [2]

54. A student makes a simple pendulum by attaching a mass to the free end of a 1.50-meter length of string suspended from the ceiling of her physics classroom. She pulls the mass up to her chin and releases it from rest, allowing the pendulum to swing in its curved path. Her classmates are surprised that the mass doesn't reach her chin on the return swing, even though she does not move. Explain why the mass does *not* have enough energy to return to its starting position and hit the girl on the chin. [1]

_____

_____

_____

55. A 6-ohm resistor and a 4-ohm resistor are connected in series with a 6-volt battery in an operating electric circuit. A voltmeter is connected to measure the potential difference across the 6-ohm resistor. In the space below, draw a diagram of this circuit including the battery, resistors, and voltmeter using symbols from the *Reference Tables for Physical Setting/Physics*. Label each resistor with its value. [Assume the availability of any number of wires of negligible resistance.] [2]

56. When a spring is compressed 2.50×10^{-2} meter from its equilibrium position, the total potential energy stored in the spring is 1.25×10^{-2} joule. Calculate the spring constant of the spring. [Show all work, including the equation and substitution with units.] [2]

Base your answers to questions 57 and 58 on the information below.

A 3.50-meter length of wire with a cross-sectional area of 3.14×10^{-6} meter^2 is at 20° Celsius. The current in the wire is 24.0 amperes when connected to a 1.50-volt source of potential difference.

57. Determine the resistance of the wire. [1]   _____ Ω

58. Calculate the resistivity of the wire. [Show all work, including the equation and substitution with units.] [2]

Base your answers to questions 59 and 60 on the information below.

In an experiment, a 0.028-kilogram rubber stopper is attached to one end of a string. A student whirls the stopper overhead in a horizontal circle with a radius of 1.0 meter. The stopper completes 10. revolutions in 10. seconds.

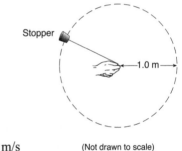

59. Determine the speed of the whirling stopper. [1]   _____ m/s   (Not drawn to scale)

60. Calculate the magnitude of the centripetal force on the whirling stopper. [Show all work, including the equation and substitution with units.] [2]

Answer all questions in this part.
Directions (61–75): Record your answers in the spaces provided.

Data Table

Base your answers to questions 61 through 64 on the accompanying information.

In a laboratory investigation, a student applied various downward forces to a vertical spring. The applied forces and the corresponding elongations of the spring from its equilibrium position are recorded in the accompanying data table.

Force (N)	Elongation (m)
0	0
0.5	0.010
1.0	0.018
1.5	0.027
2.0	0.035
2.5	0.046

June 2010

Directions (61–63): Construct a graph on the grid, following the directions below.

61. Mark an appropriate scale on the axis labeled "Force (N)." [1]

62. Plot the data points for force versus elongation. [1]

63. Draw the best-fit line or curve. [1]

64. Using your graph, calculate the spring constant of this spring. [Show all work, including the equation and substitution with units.] [2]

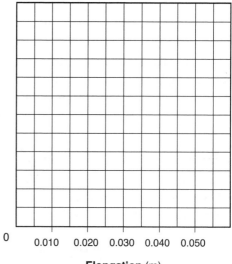

Force vs. Elongation

Base your answers to questions 65 through 68 on the information below.

An ice skater applies a horizontal force to a 20.-kilogram block on frictionless, level ice, causing the block to accelerate uniformly at 1.4 meters per second² to the right. After the skater stops pushing the block, it slides onto a region of ice that is covered with a thin layer of sand. The coefficient of kinetic friction between the block and the sand-covered ice is 0.28.

65. Calculate the magnitude of the force applied to the block by the skater. [Show all work, including the equation and substitution with units.] [2]

66. On the diagram below, starting at point *A*, draw a vector to represent the force applied to the block by the skater. Begin the vector at point *A* and use a scale of 1.0 centimeter = 5.0 newtons. [1]

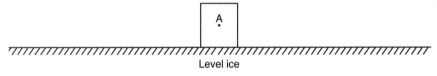

Level ice

67. Determine the magnitude of the normal force acting on the block. [1]   _____N

68. Calculate the magnitude of the force of friction acting on the block as it slides over the sand-covered ice. [Show all work, including the equation and substitution with units.] [2]

Base your answers to questions 69 through 71 on the information below and the accompanying diagram.

A monochromatic light ray ($f = 5.09 \times 10^{14}$ Hz) traveling in air is incident on the surface of a rectangular block of Lucite.

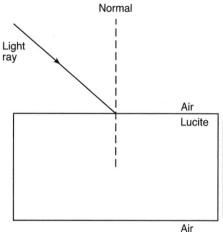

69. Measure the angle of incidence for the light ray to the *nearest degree*. [1]

_____ °

70. Calculate the angle of refraction of the light ray when it enters the Lucite block. [Show all work, including the equation and substitution with units.] [2]

71. What is the angle of refraction of the light ray as it emerges from the Lucite block back into air? [1]   _____ °

June 2010

Base your answers to questions 72 through 75 on the information below.

As a mercury atom absorbs a photon of energy, an electron in the atom changes from energy level *d* to energy level *e*.

72. Determine the energy of the
absorbed photon in electronvolts. [1]   _____ eV

73. Express the energy of the
absorbed photon in joules. [1]   _____ J

74. Calculate the frequency of the absorbed photon. [Show all work, including the equation and substitution with units.] [2]

75. Based on your calculated value of the frequency
of the absorbed photon, determine its classification
in the electromagnetic spectrum. [1]   _____

Part A
Answer all questions in this part.
Directions (1–35): For *each* statement or question, write in the space provided the *number* of the word or expression that, of those given, best completes the statement or answers the question.

1. Scalar is to vector as
(1) speed is to velocity   (3) displacement is to velocity
(2) displacement is to distance   (4) speed is to distance   1_____

2. If a car accelerates uniformly from rest to 15 meters per second over a distance of 100. meters, the magnitude of the car's acceleration is
(1) 0.15 m/s²   (2) 1.1 m/s²   (3) 2.3 m/s²   (4) 6.7 m/s²   2_____

3. An object accelerates uniformly from 3.0 meters per second east to 8.0 meters per second east in 2.0 seconds. What is the magnitude of the acceleration of the object?
(1) 2.5 m/s²   (2) 5.0 m/s²   (3) 5.5 m/s²   (4) 11 m/s²   3_____

4. A rock is dropped from a bridge. What happens to the magnitude of the acceleration and the speed of the rock as it falls? [Neglect friction.]
(1) Both acceleration and speed increase.
(2) Both acceleration and speed remain the same.
(3) Acceleration increases and speed decreases.
(4) Acceleration remains the same and speed increases.   4_____

5. A soccer ball kicked on a level field has an initial vertical velocity component of 15.0 meters per second. Assuming the ball lands at the same height from which it was kicked, what is the total time the ball is in the air? [Neglect friction.]
(1) 0.654 s   (2) 1.53 s   (3) 3.06 s   (4) 6.12 s   5_____

6. A student is standing in an elevator that is accelerating downward. The force that the student exerts on the floor of the elevator must be
(1) less than the weight of the student when at rest
(2) greater than the weight of the student when at rest
(3) less than the force of the floor on the student
(4) greater than the force of the floor on the student   6_____

7. The magnitude of the centripetal force acting on an object traveling in a horizontal, circular path will *decrease* if the
(1) radius of the path is increased
(2) mass of the object is increased
(3) direction of motion of the object is reversed
(4) speed of the object is increased   7_____

8. The centripetal force acting on the space shuttle as it orbits Earth is equal to the shuttle's
(1) inertia   (2) momentum   (3) velocity   (4) weight   8_____

9. As a box is pushed 30. meters across a horizontal floor by a constant horizontal force of 25 newtons, the kinetic energy of the box increases by 300. joules. How much total internal energy is produced during this process?
(1) 150 J   (2) 250 J   (3) 450 J   (4) 750 J   9_____

10. What is the power output of an electric motor that lifts a 2.0-kilogram block 15 meters vertically in 6.0 seconds?
(1) 5.0 J   (2) 5.0 W   (3) 49 J   (4) 49 W   10_____

11. Four identical projectiles are launched with the same initial speed, v, but at various angles above the level ground. Which diagram represents the initial velocity of the projectile that will have the largest total horizontal displacement? [Neglect air resistance.]

v 30.°	v 45°	v 60.°	v 70.°
Level ground	Level ground	Level ground	Level ground
(1)	(2)	(3)	(4) 11_____

12. Two forces act concurrently on an object on a horizontal, frictionless surface, as shown in the diagram. What additional force, when applied to the object, will establish equilibrium?

10. N → Object ← 6 N

Horizontal, frictionless surface

(1) 16 N toward the right   (3) 4 N toward the right
(2) 16 N toward the left   (4) 4 N toward the left   12_____

13. As shown in the diagram, an open box and its contents have a combined mass of 5.0 kilograms. A horizontal force of 15 newtons is required to push the box at a constant speed of 1.5 meters per second across a level surface. The inertia of the box and its contents increases if there is an increase in the

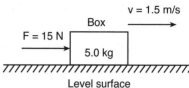

(1) speed of the box
(2) mass of the contents of the box
(3) magnitude of the horizontal force applied to the box
(4) coefficient of kinetic friction between the box and the level surface   13____

14. Which statement describes the kinetic energy and total mechanical energy of a block as it is pulled at constant speed up an incline?
(1) Kinetic energy decreases and total mechanical energy increases.
(2) Kinetic energy decreases and total mechanical energy remains the same.
(3) Kinetic energy remains the same and total mechanical energy increases.
(4) Kinetic energy remains the same and total mechanical energy remains the same.   14____

15. Which diagram represents the electric field lines between two small electrically charged spheres?

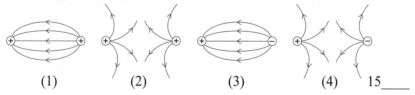

(1)   (2)   (3)   (4)   15____

16. The diagram represents a view from above of a tank of water in which parallel wave fronts are traveling toward a barrier. Which arrow represents the direction of travel for the wave fronts after being reflected from the barrier?
(1) A   (3) C
(2) B   (4) D

16____

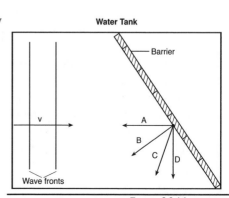

17. Two metal spheres, *A* and *B*, possess charges of 1.0 microcoulomb and 2.0 microcoulombs, respectively. In the diagram below, arrow *F* represents the electrostatic force exerted on sphere *B* by sphere *A*.

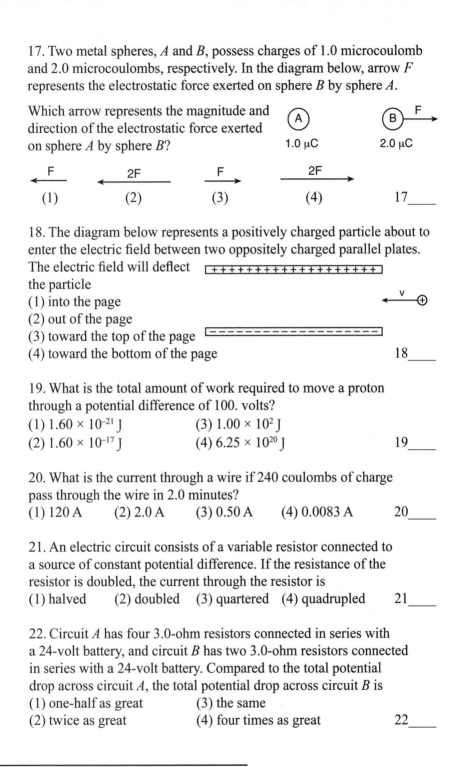

Which arrow represents the magnitude and direction of the electrostatic force exerted on sphere *A* by sphere *B*?

(A) 1.0 μC

(B) ⟶ F 2.0 μC

⟵ F
(1)

⟵ 2F
(2)

F ⟶
(3)

2F ⟶
(4)

17____

18. The diagram below represents a positively charged particle about to enter the electric field between two oppositely charged parallel plates. The electric field will deflect the particle

(1) into the page
(2) out of the page
(3) toward the top of the page
(4) toward the bottom of the page

18____

19. What is the total amount of work required to move a proton through a potential difference of 100. volts?
(1) 1.60×10^{-21} J   (3) 1.00×10^{2} J
(2) 1.60×10^{-17} J   (4) 6.25×10^{20} J   19____

20. What is the current through a wire if 240 coulombs of charge pass through the wire in 2.0 minutes?
(1) 120 A   (2) 2.0 A   (3) 0.50 A   (4) 0.0083 A   20____

21. An electric circuit consists of a variable resistor connected to a source of constant potential difference. If the resistance of the resistor is doubled, the current through the resistor is
(1) halved   (2) doubled   (3) quartered   (4) quadrupled   21____

22. Circuit *A* has four 3.0-ohm resistors connected in series with a 24-volt battery, and circuit *B* has two 3.0-ohm resistors connected in series with a 24-volt battery. Compared to the total potential drop across circuit *A*, the total potential drop across circuit *B* is
(1) one-half as great   (3) the same
(2) twice as great   (4) four times as great   22____

23. How much total energy is dissipated in 10. seconds in a
4.0-ohm resistor with a current of 0.50 ampere?
(1) 2.5 J   (2) 5.0 J   (3) 10. J   (4) 20. J   23____

24. Moving a length of copper wire through a magnetic field may
cause the wire to have a
(1) potential difference across it   (3) lower resistivity
(2) lower temperature   (4) higher resistance   24____

25. A pulse traveled the length of a stretched spring.
The pulse transferred
(1) energy, only   (3) both energy and mass
(2) mass, only   (4) neither energy nor mass   25____

26. The graph represents the
displacement of a particle in a
medium over a period of time.
The amplitude of the wave is
(1) 4.0 s   (3) 8 cm
(2) 6.0 s   (4) 4 cm

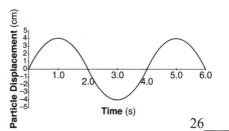

26____

27. What is the period of a water wave if 4.0 complete waves pass
a fixed point in 10. seconds?
(1) 0.25 s   (2) 0.40 s   (3) 2.5 s   (4) 4.0 s   27____

28. The diagram represents a
periodic wave. Which point
on the wave is 90° out of
phase with point P?
(1) A   (3) C
(2) B   (4) D

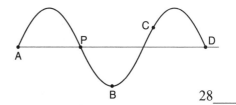

28____

29. What is the wavelength of a 256-hertz sound wave in air at STP?
(1) 1.17 × 10⁶ m   (3) 0.773 m
(2) 1.29 m   (4) 8.53 × 10⁻⁷ m   29____

30. What is the minimum total energy released when an electron
and its antiparticle (positron) annihilate each other?
(1) 1.64 × 10⁻¹³ J   (3) 5.47 × 10⁻²² J
(2) 8.20 × 10⁻¹⁴ J   (4) 2.73 × 10⁻²² J   30____

31. Which statement correctly describes one characteristic of a sound wave?
(1) A sound wave can travel through a vacuum.
(2) A sound wave is a transverse wave.
(3) The amount of energy a sound wave transmits is directly related to the wave's amplitude.
(4) The amount of energy a sound wave transmits is inversely related to the wave's frequency.   31____

32. A 256-hertz vibrating tuning fork is brought near a nonvibrating 256-hertz tuning fork. The second tuning fork begins to vibrate. Which phenomenon causes the nonvibrating tuning fork to begin to vibrate?
(1) resistance   (2) resonance   (3) refraction   (4) reflection   32____

33. Astronauts traveling toward Earth in a fastmoving spacecraft receive a radio signal from an antenna on Earth. Compared to the frequency and wavelength of the radio signal emitted from the antenna, the radio signal received by the astronauts has a
(1) lower frequency and a shorter wavelength
(2) lower frequency and a longer wavelength
(3) higher frequency and a shorter wavelength
(4) higher frequency and a longer wavelength   33____

34. On the atomic level, energy and matter exhibit the characteristics of
(1) particles, only   (3) neither particles nor waves
(2) waves, only   (4) both particles and waves   34____

35. Which particles are *not* affected by the strong force?
(1) hadrons   (2) protons   (3) neutrons   (4) electrons   35____

Part B–1
Answer all questions in this part.
Directions (36–50): For *each* statement or question, write in the space provided the *number* of the word or expression that, of those given, best completes the statement or answers the question.

36. What is the approximate diameter of an inflated basketball?
(1) 2×10^{-2} m (2) 2×10^{-1} m   (3) 2×10^{0} m   (4) 2×10^{1} m   36____

37. The graph shows the relationship between the speed and elapsed time for an object falling freely from rest near the surface of a planet. What is the total distance the object falls during the first 3.0 seconds?

(1) 12 m   (3) 44 m
(2) 24 m   (4) 72 m

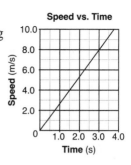

Speed vs. Time

37_____

38. A 75-kilogram hockey player is skating across the ice at a speed of 6.0 meters per second. What is the magnitude of the average force required to stop the player in 0.65 second?

(1) 120 N   (2) 290 N   (3) 690 N   (4) 920 N   38_____

39. A child pulls a wagon at a constant velocity along a level sidewalk. The child does this by applying a 22-newton force to the wagon handle, which is inclined at 35° to the sidewalk as shown to the right. What is the magnitude of the force of friction on the wagon?

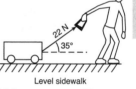

Level sidewalk

(1) 11 N   (2) 13 N   (3) 18 N   (4) 22 N   39_____

40. The diagram shows the arrangement of three small spheres, A, B, and C, having charges of $3q$, q, and q, respectively. Spheres A and C are located distance r from sphere B. Compared to the magnitude of the electrostatic force exerted by sphere B on sphere C, the magnitude of the electrostatic force exerted by sphere A on sphere C is

(1) the same   (3) $\frac{3}{4}$ as great

(2) twice as great   (4) $\frac{3}{2}$ as great

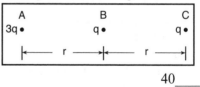

40_____

41. A space probe is launched into space from Earth's surface. Which graph represents the relationship between the magnitude of the gravitational force exerted on Earth by the space probe and the distance between the space probe and the center of Earth?

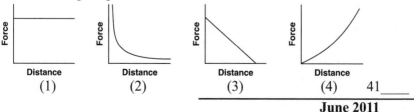

(1)   (2)   (3)   (4)   41_____

42. Which graph represents the relationship between the gravitational potential energy (*GPE*) of an object near the surface of Earth and its height above the surface of Earth?

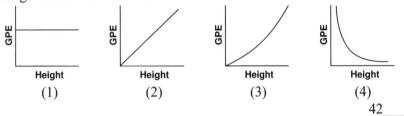

(1)   (2)   (3)   (4)

42____

43. Two parallel metal plates are connected to a variable source of potential difference. When the potential difference of the source is increased, the magnitude of the electric field strength between the plates increases. The diagram shows an electron located between the plates. Which graph represents the relationship between the magnitude of the electrostatic force on the electron and the magnitude of the electric field strength between the plates?

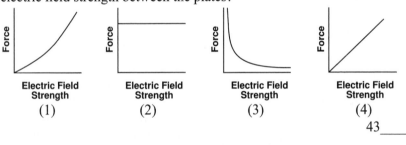

(1)   (2)   (3)   (4)

43____

44. The diagram represents a circuit consisting of two resistors connected to a source of potential difference. What is the current through the 20.-ohm resistor?

(1) 0.25 A   (2) 6.0 A   (3) 12 A   (4) 4.0 A

44____

45. The diagram below shows the magnetic field lines between two magnetic poles, *A* and *B*. Which statement describes the polarity of magnetic poles *A* and *B*?
(1) *A* is a north pole and *B* is a south pole.
(2) *A* is a south pole and *B* is a north pole.
(3) Both *A* and *B* are north poles.
(4) Both *A* and *B* are south poles.

45____

46. The diagram represents a
transverse water wave propagating
toward the left. A cork is floating
on the water's surface at point *P*.
In which direction will the cork
move as the wave passes point *P*?

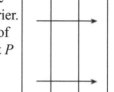

(1) up, then down, then up   (3) left, then right, then left
(2) down, then up, then down   (4) right, then left, then right   46____

47. The diagram shows a series of wave
fronts approaching an opening in a barrier.
Point *P* is located on the opposite side of
the barrier. The wave fronts reach point *P*
as a result of

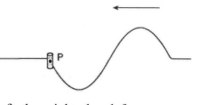

(1) resonance   (3) reflection
(2) refraction (4) diffraction   47____

48. The accompanying diagram represents
a standing wave. The number of nodes
and antinodes shown in the diagram is

(1) 4 nodes and 5 antinodes   (3) 6 nodes and 5 antinodes
(2) 5 nodes and 6 antinodes   (4) 6 nodes and 10 antinodes   48____

49. A deuterium nucleus consists of one proton and one neutron.
The quark composition of a deuterium nucleus is
(1) 2 up quarks and 2 down quarks
(2) 2 up quarks and 4 down quarks
(3) 3 up quarks and 3 down quarks
(4) 4 up quarks and 2 down quarks   49____

50. The diagram shows two
waves traveling in the same
medium. Points *A*, *B*, *C*, and
D are located along the rest
position of the medium. The
waves interfere to produce
a resultant wave. The
superposition of the waves
produces the greatest positive
displacement of the medium
from its rest position at point

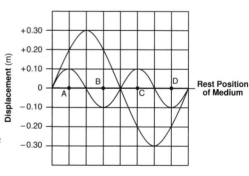

(1) *A*   (2) *B*   (3) *C*   (4) *D*   50____

Part B–2
Answer all questions in this part.
Directions **(51–65): Record your answers in the spaces provided.**

51–52. A 0.50-kilogram frog is at rest on the bank surrounding a pond of water. As the frog leaps from the bank, the magnitude of the acceleration of the frog is 3.0 meters per second^2. Calculate the magnitude of the net force exerted on the frog as it leaps. [Show all work, including the equation and substitution with units.] [2]

Base your answers to questions 53 through 55 on the information below.

A student and the waxed skis he is wearing have a combined weight of 850 newtons. The skier travels down a snow-covered hill and then glides to the east across a snow-covered, horizontal surface.

53. Determine the magnitude of the normal force exerted by the snow on the skis as the skier glides across the horizontal surface. [1]

53 _____ N

54–55. Calculate the magnitude of the force of friction acting on the skis as the skier glides across the snow-covered, horizontal surface. [Show all work, including the equation and substitution with units.] [2]

56–57. Calculate the kinetic energy of a particle with a mass of 3.34×10^{-27} kilogram and a speed of 2.89×10^5 meters per second. [Show all work, including the equation and substitution with units.] [2]

58. A simple circuit consists of a 100.-ohm resistor connected to a battery. A 25-ohm resistor is to be connected in the circuit. Determine the *smallest* equivalent resistance possible when both resistors are connected to the battery. [1]

June 2011

58 _____ Ω

59. The graph represents the relationship between the work done by a person and time. Identify the physical quantity represented by the slope of the graph. [1]

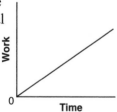

59 _____

60–61. The heating element in an automobile window has a resistance of 1.2 ohms when operated at 12 volts. Calculate the power dissipated in the heating element. [Show all work, including the equation and substitution with units.] [2]

62–63. An electromagnetic wave of wavelength 5.89×10^{-7} meter traveling through air is incident on an interface with corn oil. Calculate the wavelength of the electromagnetic wave in corn oil. [Show all work, including the equation and substitution with units.] [2]

64. The energy required to separate the 3 protons and 4 neutrons in the nucleus of a lithium atom is 39.3 megaelectronvolts. Determine the mass equivalent of this energy, in universal mass units. [1]

64 _____ u

65. A wave generator having a constant frequency produces parallel wave fronts in a tank of water of two different depths. The diagram below represents the wave fronts in the deep water.

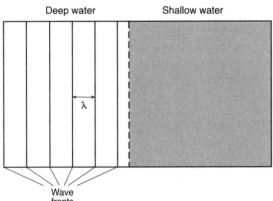

As the wave travels from the deep water into the shallow water, the speed of the waves decreases. On the diagram above, use a straightedge to draw *at least three* lines to represent the wave fronts, with appropriate spacing, in the shallow water. [1]

Part C

Answer all questions in this part.

***Directions* (66–85): Record your answers in the spaces provided.**

Base your answers to questions 66 through 69 on the information and diagram below.

A model airplane heads due east at 1.50 meters per second, while the wind blows due north at 0.70 meter per second. The scaled diagram below represents these vector quantities.

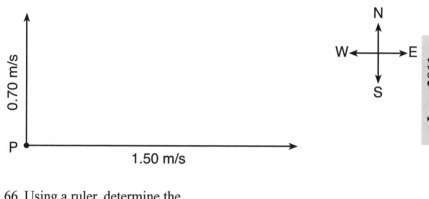

66. Using a ruler, determine the scale used in the vector diagram. [1] 1.0 cm = _____ m/s

67. On the diagram above, use a protractor and a ruler to construct a vector to represent the resultant velocity of the airplane. Label the vector *R*. [1]

68. Determine the magnitude of the resultant velocity. [1]

68_____ m/s

69. Determine the angle between north and the resultant velocity. [1]

69_____ °

Base your answers to questions 70 through 73 on the information below.

A vertically hung spring has a spring constant of 150. newtons per meter. A 2.00-kilogram mass is suspended from the spring and allowed to come to rest.

70–71. Calculate the elongation of the spring produced by the suspended 2.00-kilogram mass. [Show all work, including the equation and substitution with units.] [2]

72–73. Calculate the total elastic potential energy stored in the spring due to the suspended 2.00-kilogram mass. [Show all work, including the equation and substitution with units.] [2]

Base your answers to questions 74 through 76 on the information and diagram below.

A circuit contains a 12.0-volt battery, an ammeter, a variable resistor, and connecting wires of negligible resistance, as shown to the right.

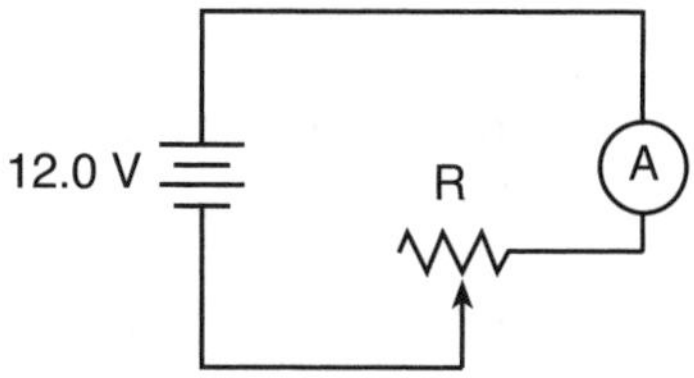

12.0 V ⎓   R   (A)

The variable resistor is a nichrome wire, maintained at 20.°C. The length of the nichrome wire may be varied from 10.0 centimeters to 90.0 centimeters. The ammeter reads 2.00 amperes when the length of the wire is 10.0 centimeters.

74. Determine the resistance of the 10.0-centimeter length of nichrome wire. [1]

74 _____ Ω

75–76. Calculate the cross-sectional area of the nichrome wire. [Show all work, including the equation and substitution with units.] [2]

Base your answers to questions 77 through 80 on the information below.

A photon with a wavelength of 2.29×10^{-7} meter strikes a mercury atom in the ground state.

77–78. Calculate the energy, in joules, of this photon. [Show all work, including the equation and substitution with units.] [2]

79. Determine the energy, in electronvolts, of this photon. [1]

79 _____ eV

80. Based on your answer to question 79, state if this photon can be absorbed by the mercury atom. Explain your answer. [1]

Base your answers to questions 81 through 85 on the information below.

A ray of monochromatic light ($f = 5.09 \times 10^{14}$ Hz) passes through air and a rectangular transparent block, as shown in the accompanying diagram.

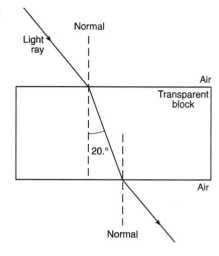

81. Using a protractor, determine the angle of incidence of the light ray as it enters the transparent block from air. [1]

_____ °

82–83. Calculate the absolute index of refraction for the medium of the transparent block. [Show all work, including the equation and substitution with units.] [2]

84–85. Calculate the speed of the light ray in the transparent block. [Show all work, including the equation and substitution with units.] [2]

June 2011

PHYSICAL SETTING

PHYSICS

ANSWERS

AND

EXPLANATIONS

A Physics Reference Table (RT) is quoted throughout this section. The Physics Reference Table can be found in the back of this booklet.

June 2008
Part A

1. 1   A vector is defined as a quantity possessing magnitude (size) and direction. Velocity is defined as a speed in a particular direction and is therefore a vector quantity. Speed is the magnitude of the velocity vector.

2. 1   Friction is a force that acts opposite to the direction of motion of an object. The air friction acting on the projectile will cause a lower and shorter trajectory.

3. 4   Inertia is a property of matter that causes an object to remain at rest or in uniform motion unless acted upon by an outside or unbalanced force. The inertia of an object is measured quantitatively by the object's mass. The larger the mass, the larger the inertia. In the RT under mechanics, find the equation $p = mv$. Momentum is the product of the mass and velocity of the object. For cart A and cart B, these products are equal to each other.

4. 3   Under Mechanics in the RT, find the equation $\bar{v} = d/t$. The speed of light is found on the List of Physical Constants in the RT. Since it is a constant, it is the average speed. The mean distance from the Earth to the Sun is also found on the List of Physical Constants. Solving for
t and substituting gives $t = \dfrac{(1.50 \times 10^{11} \text{ m})}{(3.00 \times 10^{8} \text{ m/s})} = 5.00 \times 10^{2}$ s.

5. 3   In the RT under Mechanics, find the equation $d = v_i t + \frac{1}{2} a t^2$.
Since the rock starts from rest, $v_i = 0$.
Substitution gives $0.72 \text{ m} = 0 + \frac{1}{2}(a)(0.63 \text{ s})^2$. Solving, $a = 3.6 \text{ m/s}^2$.

6. 1   In the vertical, both stones start from rest. They fall the same vertical distance and undergo the same acceleration, that due to gravity. Therefore, they both reach the ground in the same time. The horizontal velocity determines how far they travel in the horizontal.

7. 3   Under Mechanics in the RT, find the equation $a = \dfrac{\Delta v}{t}$ and $v_f^2 = v_i^2 + 2ad$.
Using the acceleration equation, $a = \dfrac{(16.0 \text{ m/s} - 8.0 \text{ m/s})}{(10. \text{ s})} = 0.80 \text{ m/s}^2$.

Substituting into the second equation gives
$(16.0 \text{ m/s})^2 = (8.0 \text{ m/s})^2 + 2(0.80 \text{ m/s}^2)(d)$. Solving, $d = 1.2 \times 10^2$ m
or Under Mechanics, find the equation $\bar{v} = d/t$. To find the average
speed, $\bar{v} = \dfrac{(v_i + v_f)}{2}$. Substituting, $\bar{v} = \dfrac{(8.0 \text{ m/s} + 16.0 \text{ m/s})}{2} = 12.0 \text{ m/s}$.
Substitution into the first equation gives $(12.0 \text{ m/s}) = \dfrac{d}{(10. \text{ s})}$.
Solving for d, $d = 1.2 \times 10^2$ m.

8. 3   In the RT under Mechanics, find the equation $g = F_g/m$. In this equation, g represents the acceleration due to gravity or the gravitational field strength. The normal force acting on the vehicle is the weight of the vehicle, F_g. Solving for F_g, $F_g = mg$.
Substituting, $F_g = (1200 \text{ kg})(4.8 \text{ m/s}^2) = 4400 \text{ N}$.

9. 2   When an object is in equilibrium, the net force acting on the object is 0. In choice 2, the sum of the horizontal forces is 0 and the sum of the vertical forces is 0. Therefore, the net force acting on the object is 0.

10. 2   In circular motion, the velocity vector is tangent to the circle and the centripetal force acts toward the center of the circle. In the position shown in the diagram, the velocity is up and the centripetal force is to the left.

11. 2   Under Mechanics in the RT, find the equation $A_x = A \cos\theta$. Since the plane is flying southeast, the eastward component will be along the x direction. Substituting and solving,
$A_x = (750. \text{ km/h})(\cos 30.0°) = 650. \text{ km/h}$.

12. 3   In the RT under Mechanics, find the equation $F_f = \mu F_N$. Open to the table of Approximate Coefficients of Friction. Since the skier is in motion, the kinetic coefficient of friction between waxed ski and snow must be used. The normal force is the weight of the skier. Use the equation $g = F_g/m$ to find the weight of the skier. The value of g is on the List of Physical Constants in the RT. Solving, $F_g = mg = (80 \text{kg})(9.81 \text{ m/s}^2) = 784 \text{N}$. Solving for the force of friction, $F_f = (0.05)(784 \text{ N}) = 40 \text{ N}$. Since the skier is moving with a constant velocity, the force pushing him forward must be equal to the opposing force of friction.

13. 3   Under Mechanics in the RT, find the equations $F_c = ma_c$ and $a_c = v^2/r$. Substituting into the first equation for a_c gives $F_c = mv^2/r$. Substituting into this equation and solving gives $F_c = \dfrac{(1750 \text{ kg})(15.0 \text{ m/s})^2}{(45.0 \text{ m})} = 8750 \text{ N}$.

14. 2   In the RT under Mechanics, find the equations $J = \Delta p$ and $p = mv$. Since the mass of an object is constant, a change in momentum is caused by a change in the object's velocity. Therefore, the second equation may be written as $\Delta p = m\Delta v$ and the first equation may now be written as $J = m\Delta v$. The football goes from a speed of 22 m/s to rest, therefore the Δv is 22 m/s. Substituting and solving gives $J = (0.45 \text{ kg})(22 \text{ m/s}) = 9.9 \text{ N•s}$. This change in momentum (impulse) is imparted to the receiver. The mass of the receiver does not matter.

15. 3   Newton's Third Law states that for every action, there is an equal and opposite reaction. The magnitude of the force the hammer exerts on the nail and the magnitude of the force that the nail exerts on the hammer are equal.

16. 3   Under Mechanics in the RT, find the equation $F_g = Gm_1m_2/r^2$. This equation indicates that the gravitational force between two objects varies inversely with the square of the distance between their centers. In moving from 16 Earth radii to 2 Earth radii from the center of the Earth, the distance has become 1/8 as great. Squaring 1/8 gives 1/64 and inverting shows that the force has become 64 times as great. See $4d$ in Helpful Hints for Physics in the back of this book.

17. 1   The work done against gravity by the student will be equal to the increase in potential energy as the student climbs the ladder. In the RT under Mechanics, find the equation $\Delta PE = mg\Delta h$. The value of g is found on the List of Physical Constants in the RT.
Substitution gives $\Delta PE = W = (60.\ \text{kg})(9.81\ \text{m/s}^2)(4.0\ \text{m})$.
Solving gives 2.4×10^3 J. The time has no effect on the work done by the student.

18. 4   Under Mechanics in the RT, find the equation $P = W/t$. Solving for W, $W = Pt$. Substitution into this equation and solving gives $W = (6000.\ W)(10.\ \text{s}) = 6.0 \times 10^4$ J.

19. 2   In the RT under Mechanics, find the equations $\Delta PE = mg\Delta h$ and $KE = \frac{1}{2}mv^2$. As the car travels up the hill, Δh increases, therefore increasing the potential energy of the car. Since the car is moving at a constant speed, the kinetic energy of the car remains the same.

20. 1   In the RT under Electricity - Parallel Circuits, find the equation $\frac{1}{Req} = \frac{1}{R_1} + \frac{1}{R_2} + \frac{1}{R_3} + ...$. Because this is a reciprocal relationship, the equivalent resistance of the circuit must be smaller than the smallest resistance.

21. 1   Under Electricity in the RT, find the equation $P = V^2/R$. This indicates that the power dissipated varies inversely with the resistance. Therefore, as the resistance increases, the power decreases. See $4c$ in Helpful Hints for Physics in the back of this book.

22. 1   Since the electron is attracted to plate A, plate A must be positively charged. By definition, the direction of an electric field is from positive to negative. Therefore, the electric field is directed from plate A toward plate B.

23. 4   In the RT under Electricity, find the equation $V = W/q$. Substitution gives $(10.0 \text{ V}) = (2.0 \times 10^{-2} \text{ J})/(q)$. Solving, $q = 2.0 \times 10^{-3} \text{ C}$.

24. 1   Under Electricity in the RT, find the equation $R = V/I$. Rearranging gives $I = V/R$. If the temperature remains constant, the resistance remains constant. Since this shows a direct relationship, as the voltage increases, the current will increase. See 4*a* in Helpful Hints for Physics in the back of this book.

25. 2   By definition, the time required for a wave to complete one cycle is the period of the wave.

26. 3   Open to the Electromagnetic Spectrum chart in the RT. The AM-band radio waves have a wavelength range of 10^2 to 10^3 m.

27. 2   The wavelength of a wave is defined as the distance between two consecutive points on a wave that are in phase. Two points are said to be in phase when they have the same displacement from the rest or equilibrium position and are moving in the same direction. Points B and F are the only two points that meet this condition.

28. 4   Refraction is defined as the change in direction of travel of a wave as it obliquely enters a medium in which its speed changes. As a wave enters a new medium, the color, frequency and period remain the same.

29. 4   In the RT under Waves, find the equation $n = c/v$. The speed of light (c) is found on the List of Physical Constants in the RT. Substitution gives $n = (3.00 \times 10^8 \text{ m/s})/(2.00 \times 10^8 \text{ m/s})$. Solving, $n = 1.50$.

30. 4   Standing waves are commonly produced by the reflection of a wave back upon itself from a fixed barrier. These two waves therefore will have the same frequency, amplitude and be traveling in opposite directions.

31. 4   The Doppler Effect is the change in frequency of a wave produced by relative motion between the source of the wave and the receiver. If the distance between the source and receiver decreases, the frequency received increases. If the distance between the source and receiver increases, the frequency decreases. Since the car is moving toward the observer (receiver), the frequency is higher than the actual frequency.

32. 2   When two waves encounter each other in the same medium, each moves as if the other was not present. Therefore, after the encounter, each continues to travel in the same direction with the same shape. During the encounter, they simply interfer with each other.

33. 2   In the RT, find the Energy Level Diagrams. Referring to that of Mercury, the ionization energy is 10.38 eV, the difference between the ground state (Level a) and the ionization state (Level j). The excess energy carried by the photon will be carried away by the electron as kinetic energy (20.00 eV – 10.38 eV = 9.62 eV).

34. 4   This force is the force of attraction between oppositely charged particles. Therefore, it is electromagnetic force.

35. 4   Under Modern Physics in the RT, find the equation $E = mc^2$. The speed of light (c) is found on the List of Physical Constants in the RT. Substitution and solving gives $E = (1.00 \text{ kg})(3.00 \times 10^8 \text{ m/s})^2 = 9.00 \times 10^{16} \text{ J}$.

Part B-1

36. 4   Distance is a scalar quantity and therefore has no direction associated with it. It is the magnitude of the displacement vector. During the time 0.0 s to 4.0 s, the distance traveled is 8 m. From 4.0 s to 6.0 s, the object was at rest and traveled 0 m. In the interval 6.0 s to 8.0 s, the object traveled 8 m. In the last interval, 8.0 s to 10.0 s, the traveled 8 m. The object total distance is 8 m + 0 m + 8 m + 8 m = 24 m.

37. 3   To answer this question, you must be familiar with the standard kilogram. This experience you probably got in the physics lab or class. Using the Prefixes for Powers of 10 chart in the RT, choice 1 is one million kg, choice 2 is one thousand kg, choice three is one thousandth of a kg and choice 4 is one millionth of a kg. Choice 3 is the only choice that is reasonable.

38. 3   The gravitational force between two objects is one of attraction. The Earth and satellite attract each other with equal but opposite forces. The force of the Earth on the satellite is directed toward the Earth and the force of the satellite on the Earth is directed toward the satellite. Choice 3 shows the proper directions for these forces.

39. 2   Since the block is moving with a constant velocity, the force F must equal the frictional force plus the component of the blocks weight that acts down the incline, parallel to the frictional force. The weight of the object, which acts straight down, may be resolved into two components, one acting parallel to the incline, and the other perpendicular to the incline. In the force triangle formed by the weight, the parallel component and the perpendicular component, the parallel component is the side opposite the 30° angle in the force triangle. This component is 5.0 N (half the weight, which is the hypotenuse). Force F is therefore 3.0 N + 5.0 N = 10.0 N.

40. 1   Under Mechanics in the RT, find the equation $a = F_{net}/m$. The net force acting on the object is the difference between the oppositely directed 25 N and 35 N forces, or 10. N. Substitution into the equation gives $a = (10. \text{ N})/(15 \text{ kg})$. Solving, $a = 0.67 \text{ m/s}^2$.

41. 3   The force that will produce equilibrium with the original two forces must have a magnitude equal to and a direction opposite the resultant of the two forces. It is called the equilibrant of the two forces. The resultant of the two forces has a direction that is between that of the two forces. Therefore, the force in choice 3 will produce equilibrium.

42. 2   In the RT under Mechanics, find the equation $a_c = v^2/r$. This shows that the centripetal acceleration varies directly with the square of the speed. See 4b under Helpful Hints for Physics in the back of this book.

43. 4   Under Mechanics in the RT, find the equations $p = mv$ and $p_{before} = p_{after}$. Before the collision, the momentum of mass m_A is $m_A v$ and that of mass m_B is 0 since it is at rest. After the collision, the momentum of mass m_A is $m_A v'$ and that of mass m_B is $m_B v'$. Substitution into the second equation gives $m_A + 0 = m_A v' + m_B v'$. Solving for v': $v' = (m_A v)/(m_A + m_B)$.

44. 4   Energy, and its equivalent, work, are measured in joules (J). In the RT under Mechanics, find the equation $KE = \frac{1}{2} mv^2$. The unit of kinetic energy is the unit of mass (kg) multiplied by the square of the unit of speed (m/s). In fundamental units, energy is expressed in kg $\cdot$ m^2/s^2.

45. 1   As an object rises into the air, it speed decreases. Therefore, its kinetic energy decreases. As it rises into the air, the distance from the surface of the Earth increases, thereby increasing its potential energy. The graphs in choice 1 show these changes.

46. 3   In the RT under Electricity, find the equation $I = \Delta q/t$. In this equation, Δq must be in coulombs (C). In the List of Physical Constants in the RT, $1 \text{ C} = 6.25 \times 10^{18} \text{ e}$. Using this as a conversion factor,
$$2.50 \times 10^{16} \text{ e} \times \frac{1 \text{ C}}{6.25 \times 10^{18} \text{ e}} = 0.004 \text{ C}.$$

Substituting and solving, $\quad I = \dfrac{(0.004 \text{ C})}{(1s)} = 4.00 \times 10^{-3} \text{ A}$

47. 1   Work and energy are equivalent quantities. Under Electricity in the RT, find the equation $W = VIt$. In this equation, time must be in seconds. Substituting and solving, $W = E = (120. \text{ V})(3.00 \text{ A})(60.0 \text{ s}) = 2.16 \times 10^4 \text{ J}$.

48. 2   Point A moves up and down as the wave passes. It does not move along with the wave. As the wave moves to the right, the bottom of a trough is approaching point A. Therefore, point A is moving down.

49. 3   Under Modern Physics in the RT, find the equation $E_{photon} = hf$. This shows that the energy of the photon varies directly with the frequency. See 4a under Helpful Hints for Physics in the back of this book.

50. 4   The charge and mass of a proton, as given on the table, are +1 and 0.938 GeV/c^2, respectively. The particle in question must have a charge of -1 and a mass greater than 0.938 GeV/c^2. This is the omega particle.

51. 2   The table indicates that all of the particles listed on the table are composed of quarks. Find the Classification of Matter chart in the RT. The hadrons are composed of quarks.

Part B-2

52. 50. m

Explanation: In the RT under Mechanics, find the equation $\bar{v} = d/t$. Solving for d, $d = \bar{v}t$. The average velocity can be found using $\bar{v} = \dfrac{(v_i + v_f)}{2}$. During time 0 to 2.0 s, $\bar{v} = 5$ m/s. The displacement during this interval is $d = (5$ m/s$)(2.0$ s$) = 10.$ m. From 2.0 s to 6.0 s, $\bar{v} = 10$ m/s. The displacement during this interval is $d = (10.$ m/s$)(4.0$ s$) = 40.$ m. The total displacement is $10.$ m $+ 40.$ m $= 50.$ m.

53. $KE = \Delta PE = mg\Delta h$ Explanation: In the absence of friction, the

$KE = (65$ kg$)(9.81$ m/s$^2)(5.5$ m$)$ change in kinetic energy is equal to the

$KE = 3.5 \times 10^3$ change in potential energy. The minimum kinetic energy needed must be equal to the potential energy of the vaulter at the highest point. In the RT under Mechanics, find the equations $KE = \frac{1}{2} mv^2$ and $\Delta PE = mg\Delta h$. Combining these equations, $KE = \Delta PE = mg\Delta h$. The value of g is found on the List of Physical Constants in the RT. Substitution and solving gives

$KE = \Delta PE = (65$ kg$)(9.81$ m/s$^2)(5.5$ m$) = 3503.5$ J $= 3500$ J.

54. $KE = \dfrac{1}{2}mv^2$ Explanation: Under Mechanics in the RT, find the equation $KE = \frac{1}{2} mv^2$. Using the value

$v = \sqrt{\dfrac{2KE}{m}}$ for kinetic energy calculated in question 53, 3500 J $= \frac{1}{2} (65$ kg$)(v)^2$. Solving, $v = 10.$ m/s.

$v = \sqrt{\dfrac{2(3.5 \times 10^3)}{65 \text{ kg}}}$

$v = 10.$ m/s

55. Answer: 1.0 cm = 2.0 m ± 0.2 m

Explanation: The length of the lines representing the 8.0 m and 6.0 m displacements are 4 cm and 3 cm, respectively. Therefore, the scale is 1 cm = 2 m.

56.

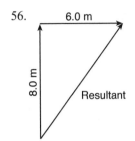

Explanation: The resultant is represented by the line drawn from the tail of the 8.0 m displacement to the head of the 6.0 m displacement.

57. Answer: 10. m ± 0.4 m

Explanation: Since the diagram is drawn to scale, measure the length of the line representing the resultant and interpret using the scale determined in question 55. The length of the line representing the resultant is 5 cm. Therefore, the displacement is 10 m.

58. $F_e = \dfrac{kq_1q_2}{r^2}$

$F_e = \dfrac{\left(8.99 \times 10^9 \text{ N} \bullet \text{m}^2/\text{C}^2\right)\left(2.0 \times 10^{-6} \text{ C}\right)\left(2.0 \times 10^{-6} \text{ C}\right)}{\left(2.0 \times 10^{-1} \text{ m}\right)^2}$

$F_e = 9.0 \times 10^{-1}$ N

Explanation: In the RT under Electricity, find the equation $F_e = \dfrac{kq_1q_2}{r^2}$.

The value of k is found on the List of Physical Constants in the RT.

Substitution gives $F_e = \dfrac{(8.99 \times 10^9 \text{ N} \bullet \text{m}^2/\text{C}^2)(2.0 \times 10^{-6} \text{ C})(2.0 \times 10^{-6} \text{ C})}{(2.0 \times 10^{-1} \text{ m})^2}$

Solving, $F_e = 9.0 \times 10^{-1}$ N.

59. Answer: 3.1×10^{-6} m^2

Explanation: In the RT under Geometry and Trigonometry, find the equation $A = \pi r^2$. Substitution gives $A = (3.14)(1.0 \times 10^{-6} \text{ m})^2$.
Solving, $A = 3.1 \times 10^{-6}$ m^2.

60.
$$R = \frac{\rho L}{A}$$

$$R = \frac{(1.72 \times 10^{-8}\,\Omega \cdot m)(10.0\,m)}{3.1 \times 10^{-6}\,m^2}$$

$$R = 5.5 \times 10^{-2}\,\Omega$$

Explanation: Under Electricity in the RT, find the equation $R = \rho L/A$. The value of ρ is found on the table of Resistivities at 20°C in the RT. Use the value for A determined in question 59. Substituting gives
$$R = \frac{(1.72 \times 10^{-8}\,\Omega \cdot m)(10.0\,m)}{(3.1 \times 10^{-6}\,m^2)}.$$
Solving, $R = 5.5 \times 10^{-2}\,\Omega$.

61.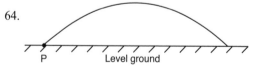

Explanation: To be 180° out of phase with point A, point **X** must be located one-half wavelength from point A. It must be an equal distance below the equilibrium position and moving in the opposite direction relative to point A.

62. $A_y = A \sin\theta$

$v_{iy} = (25\ m/s)(\sin 40.°)$

$v_{iy} = 16\ m/s$

Explanation: Under Mechanics in the RT, find the equation $A_y = A \sin\theta$. A_y represents the vertical component of the initial velocity of the ball. Substituting and solving gives
$A_y = (25\ m/s)(\sin 40°) = (25\ m/s)(0.64) = 16\ m/s$.

63. $v_f^2 = v_i^2 + 2ad$

$$d = \frac{v_f^2 - v_i^2}{2a}$$

$$d = \frac{(16\ m/s)^2}{2(9.81\ m/s^2)}$$

$d = 13\ m$

Explanation: In the RT under Mechanics, find the equation $v_f^2 = v_i^2 + 2ad$. In the vertical, $v_i = 16$ m/s (your value of the vertical component calculated in question 62) and $v_f = 0$, the speed at the highest point. The acceleration is that due to gravity (g), found on the List of Physical Constants in the RT. Substitution gives $0 = (16\,m/s)^2 + 2(9.81\,m/s^2)(d)$. Solving, $d = 13\,m$.

64.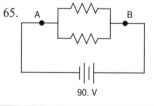

P   Level ground

Explanation: In the absence of friction, the path of the projectile is a parabola.

65. A [resistors] B

Explanation: When connected in parallel, the resistors provide separate paths for current.

90. V

66. Answer: 90.0 V

Explanation: In the RT under Electricity - Parallel Circuits, find the equation $V = V_1 = V_2 = V_3 = ...$. This equation indicates that the potential difference across each resistor is equal to the potential difference of the source (90. V).

67. $R = \dfrac{V}{I}$

$I = \dfrac{V}{R}$

$I = \dfrac{90.\ V}{15\ \Omega}$

$I = 6.0\ A$

Explanation: Under Electricity in the RT, find the equation $R = V/I$. Using the potential difference across R_1 from question 66, $15\Omega = (90.\ V)/I$. Solving for I, $I = 6.0\ A$.

68-69   **Dart's Maximum Vertical Displacement vs. Spring Compression**

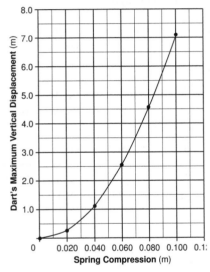

68. Explanation: Each point must be plotted to ± 0.3 grid space.

69. Explanation: Connect the points plotted on the graph in question 68 with a smooth curve for best fit.

70. $PE_s = \dfrac{1}{2}kx^2$

$PE_s = \dfrac{1}{2}(140\ N/m)(0.070\ m)^2$

$PE_s = 0.34$

Explanation: Under Mechanics in the RT, find the equation $PE_s = \dfrac{1}{2}kx^2$. Using the graph from question 69, to achieve a maximum vertical displacement of 3.5 m, the spring must be compressed 0.07 m. Substituting and solving, $PEs = \dfrac{1}{2}(140\ N/m)(0.07\ m)^2 = 0.343\ J$.

71. Answer: 5.6 N

Explanation: In the RT under Mechanics, find the equation $F_s = kx$. Substituting and solving, $F_s = (140\ N/m)(0.04\ m) = 5.6\ N$.

72. $n_1 \sin \theta_1 = n_2 \sin \theta_2$ Explanation: Under Waves in the RT, find the equation
$\sin \theta_2 = \dfrac{n_1 \sin \theta_1}{n_2}$ $n_1 \sin \theta_1 = n_2 \sin \theta_2$ (Snell's Law). The indices of refraction
   of air and corn oil are found on the table of Absolute
$\sin \theta_2 = \dfrac{1.00 \sin 35°}{1.47}$ Indices of Refraction in the RT. Substitution gives
   $(1.00)(\sin 35°) = (1.47)(\sin \theta_2)$. Solving, $\sin \theta_2 = 0.390$
$\theta_2 = 23°$ and $\theta_2 = 23°$.

73. Examples of acceptable responses:
The light does not bend because light travels at the same speed in both layers.
or   The absolute indices of refraction are the same.

Explanation: In order for a light ray to undergo refraction, the ray must enter a medium and undergo a change in speed. The index of refraction of corn oil and glycerol are the same (1.47). Therefore, the speed of the ray does not change when entering glycerol from corn oil.

74.

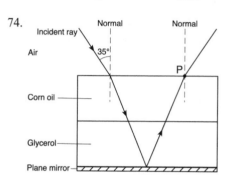

Explanation: Since all surfaces are parallel to each other, the angle of refraction in air must equal the original angle of incidence in air. Therefore, the angle of incidence in air as the ray leaves the corn oil must be 35°.

or   From the geometry of the rays, the angle of incidence at the corn oil - air surface is 23°. Using Snell's Law, $(1.47)(\sin 23°) = (1.00)(\sin \theta_2)$. $\sin \theta_2 = 0.57$ and $\theta_2 = 35°$.

75. Answer: 0.01863 u

Explanation: Using the masses from the Data Table and the information in the nuclear equation: $(1.00783 \text{ u} + 7.01600 \text{ u}) - 2(4.00260 \text{ u}) = 0.01863$ u.

76. Answer: 17.3 MeV

Explanation: The difference in mass calculated in question 75 is converted into energy. The relationship between the universal mass unit (u) and energy in megaelectron volts (MeV) is given on the List of Physical Constants in the RT ($1 \text{ u} = 9.31 \times 10^2 \text{ MeV}$). Use this as a conversion factor to change the mass difference into energy: $0.01863 \text{ u} \times \dfrac{9.31 \times 10^2 \text{ MeV}}{1 \text{ u}} = 17.3 \text{ MeV}$.

1. 3   Under Mechanics in the RT, find the equation $\bar{v}=\dfrac{d}{t}$. This states that the average speed of an object is the total distance divided by the total time. The total distance is 170 km (80. km + 50. km + 40. km) and the total time is 2.00 h (1.00 h + 0.50 h + 0.50 h). Substituting and solving gives $\bar{v} = (170.\ \text{km})/(2.00\ \text{h}) = 85\ \text{km/h}$.

2. 4   In the RT under Mechanics, find the equations $A_x = A\cos\theta$ and $A_y = A\sin\theta$. These equations give the horizontal and vertical components, respectively, of any vector A. Substitution gives $A_x = (24\ \text{N})(\cos 35°)$ and $A_y = (24\ \text{N})(\sin 35°)$. Solving, $A_x = 20.\ \text{N}$ and $A_y = 14\ \text{N}$.

3. 1   A vector quantity has both magnitude (size) and direction. Of the choices given, only impulse has both magnitude and direction.

4. 4   Under Mechanics in the RT, find the equation $\bar{v}=\dfrac{d}{t}$. Convert the distance from km to m: 300 km × (10^3 m/1 km) = 3.00×10^5 m. Substitution gives $\bar{v} =(3.00 \times 10^5\ \text{m})/(3.60 \times 10^3\ \text{s})$. Solving, $\bar{v} =83.3\ \text{m/s}$.

5. 3   Under Mechanics in the RT, find the equation $d = v_it + \frac{1}{2}at^2$. The object starts from rest, therefore $v_i = 0$. The acceleration of a freely falling object is that due to gravity (g).The value of g is on the List of Physical Constants. Substituting gives $d = 0 + \frac{1}{2}(9.81\ \text{m/s}^2)(1.0\ \text{s})^2$. Solving, $d = 4.9\ \text{m}$.

6. 2   A launch angle of 45° produces the greatest horizontal distance (range) traveled for a projectile.

7. 2   In the RT under Mechanics, find the equations $F_c = ma_c$ and $a_c = v^2/r$. Substitution of the second equation into the first gives $F_c = mv^2/r$. Substitution into this equation gives 1200. N = (200. kg)(v^2)/(25 m). Solving, $v = 12\ \text{m/s}$.

8. 3   From the equation $F_c = mv^2/r$ (see question 7), the maximum centripetal force varies inversely with the radius of the track and directly with the mass and square of the speed of the go-cart. Since 1200. N is the maximum centripetal force available, increasing the radius of the track would enable the go-cart to travel around the track at a greater speed. See 4 under Helpful Hints for Physics in the back of this book.

9. 4   The mass of an object is a quantitative measure of the inertia of an object. Since the mass of the cart remains constant, the inertia of the cart is unchanged.

10. 2   In the RT under Mechanics, find the equation $a = F_{net}/m$. The net force acting on the block is 10 N (12 N – 2 N). Substitution gives 2.0 m/s^2 = (10 N)/m. Solving, m = 5 kg.

11. 3　By definition, a body in equilibrium is at rest or moving with a constant velocity. A car moving with a constant speed along a straight, level road is moving with a constant velocity.

12. 4　Under Mechanics in the RT, find the equation $g = F_g/m$. The value of g is found on the List of Physical Constants. Substitution gives 9.81 m/s^2 = $(F_g)/(2.00$ kg$)$. Solving, $F_g = 19.6$ N.

13. 2　In the RT under Mechanics, find the equation $P = F\bar{v}$. Substitution gives 210 W = $(F)(7.0$ m/s$)$. Solving, $F = 30.$ N.

14. 3　Potential energy is energy an object possesses due to its condition or position. Under Mechanics in the RT, find the equation $\Delta PE = mg\Delta h$. The gravitational potential energy of an object depends upon it position with respect to the Earth.

15. 3　In the RT under Mechanics, find the equation $E_T = PE + KE + Q$. In free fall, the effect of friction is neglected, therefore $Q = 0$. The mechanical energy of an object is the sum of its potential and kinetic energies. This remains constant in the absence of friction.

16. 1　Under Mechanics in the RT, find the equation $F_s = kx$. Use this equation to determine x. Substitution gives 1.2 N = $(4.0$ N/m$)(x)$. Solving, $x = 0.3$ m. Now find the equation $PE_s = \frac{1}{2}kx^2$. Substituting and solving, $PE_s = \frac{1}{2}(4.0$ N/m$)(0.3$ m$)^2 = 0.18$ J.

17. 1　In the RT under Mechanics, find the equation $F = Gm_1m_2/r^2$ and under Electricity, find the equation $F_e = kq_1q_2/r^2$. Both of these equations indicate that the force involved varies inversely with the square of the distance between the centers of the objects. The distance between the centers is tripled (1.0 m to 3.0 m). The forces are then reduced to 1/9 of the original value. See 4d under Helpful Hints for Physics in this book.

18. 3　Under Electricity in the RT, find the equation $R = \rho L/A$. This equation tells us that electrical resistance of a conductor varies inversely with the cross-sectional area of the conductor. See 4c under Helpful Hints for Physics in the back of this book.

19. 2　In the RT under Electricity, find the equation R = V/I. Substitution gives $24\Omega = (6.0$ V$)/(I)$. Solving, I = 0.25 A.

20. 3　In the RT under Electricity, find the equation W = Pt. Solving for P, P = W/t. Work and energy are equivalent quantities. Substitution gives $100.$ W = W/($60.$ s). Solving, W = 6.0×10^3 J.

21. 4　Electrons are negatively charges particles. Therefore, they will be repelled by the bottom plate and attracted by the top plate.

22. 1   The gravitational force is always one of attraction. Since magnet A remains suspended above magnet B, the magnetic force must be one of repulsion.

23. 2   In the RT, find The Electromagnetic Spectrum chart. The frequency range of the colors in the visible part of the spectrum is given. Under Waves, find the equation $v = f\lambda$. For electromagnetic waves, the speed is that of light (c), which is found on the List of Physical Constants. Substitution gives $(3.00 \times 10^8 \text{ m/s}) = (f)(5.0 \times 10^{-7} \text{ m})$. Solving, $f = 6.0 \times 10^{14}$ Hz. This is in the green portion of the visible spectrum.

24. 1   Radio, television and X-rays are electromagnetic waves (see The Electromagnetic Spectrum in the RT). Electromagnetic waves do not need a medium for transmission. Sound waves, being mechanical waves, require a medium for transmission.

25. 2   The amplitude of a wave is a measure of the energy carried by the wave. If the tuning fork is struck harder, the wave will carry more energy.

26. 2   The distance between successive compressions on a wave is the wave-length of the wave. In the RT under Waves, find the equation $v = f\lambda$. The speed of sound at STP is found on the List of Physical Constants. Substitution gives $(3.31 \times 10^2 \text{ m/s}) = (440 \text{ Hz})(\lambda)$. Solving, $\lambda = 0.75$ m.

27. 3   The angle of incidence for the light ray is the angle between the incident ray and the normal, which is 60°. The angle of reflection is the angle between the reflected ray and the normal, and is equal to the angle of incidence, and therefore is also 60°. The angle between the incident light ray and the reflected ray is therefore 120°.

28. 4   Find the table of Absolute Indices of Refraction in the RT. The index of diamond is 2.42. That of crown glass is 1.52. Under Waves in the RT, find the equation $n = c/v$. Solving for v, $v = c/n$. The value of c is a constant and is found on the List of Physical Constants. Since the light is entering a medium with a lower index of refraction, as shown by the equation for v, the speed of the light will increase.

29. 2   The Principle of Superposition states that when two waves meet, the resultant wave or disturbance is the algebraic sum of the individual waves or disturbances. The two waves traveling toward each other have positive amplitudes of 5 cm and 10. cm. When they meet, the algebraic sum of the amplitudes is 15 cm.

30. 1   Resonance is the response of an object to impressed vibrations of the same frequency as the natural frequency of vibration of the object. The sound wave must have a frequency equal to the natural frequency of the glass, causing it to vibrate enough to shatter the glass.

31. 4 The charge of the alpha particle is due to the two protons, each possessing a unit positive charge. The magnitude of this charge is the same as that on an electron or the elementary charge (e), which is found on the RT. The charge on an alpha particle is therefore 2e or $2 \times (1.60 \times 10^{-19} \text{ C}) = 3.20 \times 10^{-19} \text{ C}$.

32. 4 The values of the mercury energy levels are given on the Energy Level Diagrams in the RT. In returning to the ground state (a level) from the c level of the mercury atom, the electron could go from the c level to the a level in one step or from the c level to the b level and then from the b level to the a level in two steps. Under Modern Physics in the RT, find the equation $E_{photon} = E_i - E_f$. The following photons could be emitted:

$$E_c - E_a = (-5.52 \text{ eV}) - (-10.38 \text{ eV}) = 4.86 \text{ eV}$$
$$E_c - E_b = (-5.52 \text{ eV}) - (-5.74 \text{ eV}) = 0.22 \text{ eV}$$
$$E_b - E_a = (-5.74 \text{ eV}) - (-10.38 \text{ eV}) = 4.64 \text{ eV}$$

Of the choices given, the only photon that could not be emitted is the one of 5.43 eV.

33. 2 Diffraction is the spreading of a wave into the region behind an obstacle in its path. Since only waves can undergo diffraction, the observation that light diffracts after passing through a narrow opening in a barrier is evidence that light is a wave.

34. 4 Under Mechanics in the RT, find the equation $F_g = Gm_1m_2/r^2$. This equation indicates that the force of gravity varies inversely as the square of the distance between the centers of the masses. In changing the distance from 3.84×10^8 m to 1.92×10^8 m, the distance between the centers is halved. This causes the force to become 4 times as great. Therefore, the new force is $4 \times (2.0 \times 10^{20} \text{ N}) = 8.0 \times 10^{20} \text{ N}$. See 4d under Helpful Hints for Physics in the back of this book.

35. 1 The particles of the nucleus are held together by the strong force, the strongest force in nature.

Part B-1

36. 1 This is based strictly on your familiarity of apples and the Metric System. Under Mechanics in the RT, find the equation $\Delta PE = mg\Delta h$. Work and energy are equivalent quantities. The value of g is found on the List of Physical Constants. Use 10 m/s² as an approximation to calculate the mass of the apple in each of the choices that are given using the equation above. For the energies given in the choices, the masses would be 0.1 kg, 0.001 kg, 10 kg and 100 kg, respectively. Only choice 1 is a reasonable mass of an apple. Remember that in our system, the FPS or English system, 1.00 kg = 2.20 lb.

37. 1   Acceleration is the rate of change of velocity of an object. At time $t = 5.0$ s, the car is moving with a constant velocity. The acceleration of the car is therefore zero.

38. 3   In the RT under Mechanics, find the equation $\bar{v} = d/t$. From $t = 0.0$ s to $t = 4.0$ s, the average speed is $(0.0 \text{ m/s} + 10.0 \text{ m/s})/2 = 5.0 \text{ m/s}$. The distance traveled during this time is $d = \bar{v}t = (5.0 \text{ m/s})(4.0 \text{ s}) = 20.$ m. From $t = 4.0$ s to $t = 6.0$ s, the average speed is the constant speed of 10.0 m/s. The distance traveled during this time is $d = (10.0 \text{ m/s})(2.0 \text{ s}) = 20.$ m. The total distance is then 20. m + 20. m = 40. m.

39. 2   In the RT under Mechanics, find the equation $g = F_g/m$. Using the weight of the person on Earth, find the mass of the person. The value of g at the Earth's surface in on the List of Physical constants. Substitution gives $9.81 \text{ m/s}^2 = (785 \text{ N})/m$. Solving, $m = 80.0$ kg. Since the mass of the person is constant, this is the mass of the person on Mars. Using the same equation, substitution and solving gives $g = (298 \text{ N})/(80.0 \text{ kg}) = 3.72 \text{ N/kg}$.

40. 4   Under Mechanics in the RT, find the equations $p = mv$ and $J = F_{net}t = \Delta p$. Since mass is a constant, a change in momentum is caused by a change in velocity (v). Therefore, $\Delta p = m\Delta v$. Using this in the second equation, $F_{net}t = m\Delta v$. The change in velocity of the cyclist is 20. m/s. Substitution gives $F_{net}(0.50 \text{ s}) = (60. \text{ kg})(20. \text{ m/s})$. Solving, $F_{net} = 2.4 \times 10^3$ N.

41. 3   Under Mechanics in the RT, find the equation $W = Fd$. Substituting and solving, W = (30.0 N)(4.0 m) = 120 J.

42. 3   In the RT under Mechanics, find the equations $\Delta PE = mg\Delta h$ and $KE = \frac{1}{2}mv^2$. Since the sidewalk is level and the speed is constant, the gravitational potential energy and kinetic energy of the wagon remain constant. The work done on the wagon must therefore be done against friction, causing an increase in internal energy of the wagon.

43. 1   Under Mechanics in the RT, find the equation $W = Fd$. From this equation, the unit of work would be the N•m. In the same section of the RT, find the equation $a = F_{net}/m$. Solving for F_{net}, $F_{net} = ma$. The unit of force is the kg•m/s^2. The unit of work can now be expressed as (kg•m/s^2)(m) = kg•m^2/s^2.

44. 4   In the RT under Modern Physics, find the equation $E_{photon} = hc/\lambda$. From the equation given in the question, $p = h/\lambda$, solving for h, $h = p\lambda$. Substituting for h in the equation for photon energy, $E_{photon} = (p\lambda)(c/\lambda) = pc$.

45. 1   Under Electricity in the RT, find the equation $R = V/I$. Solving for I, $I = V/R$. For a constant potential difference at constant temperature, the current (I) varies inversely with resistance (R). This relationship is shown in graph 1. See 4c under Helpful Hints for Physics in the back of this book.

46. 4   The two resistors are connected in series. Under Electricity - Series Circuits in the RT, find the equation $I = I_1 = I_2 = I_3 = \dots$. The current through the two resistors in therefore the same (4.0 A). Now find the equation $R = V/I$ under Electricity in the RT. Substitution into this equation gives $6.0\ \Omega = (V)/(4.0\ A)$. Solving, $V = 24\ V$.

47. 3   Under Electricity - Series Circuits in the RT, find the equation $R_{eq} = R_1 + R_2 + R_3 + \dots$. The resistors in 1 and 4 are in series. In choice 1, $R_{eq} = 2\ \Omega + 2\ \Omega = 4\ \Omega$ and in choice 4, $R_{eq} = 1\ \Omega + 1\ \Omega = 2\ \Omega$. Under Electricity - Parallel Circuits, find the equation $1/R_{eq} = 1/R_1 + 1/R_2 + 1/R_3 + \dots$. The resistors in 2 and 3 are in parallel. In choice 2, $1/R_{eq} = 1/(2\ \Omega) + 1/(2\ \Omega)$ and solving, $R_{eq} = 1\ \Omega$. In choice 3, $1/R_{eq} = 1/(1\ \Omega) + 1/(1\ \Omega)$, and solving, $R_{eq} = 0.5\ \Omega$.

Part B-2

48. 5.66 m

Explanation: The 4.00 m east and north displacements at are at a right angle to each other. The resultant displacement is represented by the hypotenuse of the right triangle formed by the two displacements. Under Geometry and Trigonometry - Right Triangle in the RT, find the equation $c^2 = a^2 + b^2$. The two 4.00 m displacements would be the sides a and b and the resultant is c. Substitution gives $c^2 = (4.00\ m)^2 + (4.00\ m)^2$. Solving, $c = 5.66$ m.

49. 50 N

Explanation: Newton's Third Law states that for every action, there is an equal but opposite reaction. If the puck exerts a 50 N force on the player, the player then exerts a 50 N force on the puck.

50. $P_{before}\ P_{after}$

$m_1 v_1\ m\ v_2\ _2 = 0$

$v_1 = \dfrac{-m_2 v_2}{m_1}$

$v_1 = \dfrac{-(2.5 \times 10^3\ kg)(8.0\ m/s\ )}{1.1 \times 10^3\ kg}$

$v_1 = -18$ m/s or 18 m/s

or

$m_1 v_1 = m_2 v_2$

$(1.1 \times 10^3\ kg)\ v_1 = (2.5 \times 10^3\ kg)(8.0\ m/s)$

$v_1 = 18$ m/s

Explanation: Under Mechanics in the RT, find the equations $p = mv$ and $P_{before} = P_{after}$. The total momentum before the collision is the sum of the momentums of both cars. Since they are traveling in opposite directions, one of the velocities must be given a negative sign. Let this be that of the van. Since the car and van come to rest after the collision, the total momentum after is zero. Substitute into the equation and solve for v_1.

or

Since the total momentum of the car and van after collision is zero, the momentum of the car and the momentum of the van must have been equal in magnitude to each other. Substitute into the equation and solve for v_1.

51. F_c $\dfrac{mv^2}{r}$

    v $\sqrt{\dfrac{F_c r}{m}}$

    $v = \sqrt{\dfrac{(36\,\text{N})(5.0\,\text{m})}{20.\,\text{kg}}}$

    $v = 3.0\,\text{m/s}$

Explanation: In the RT under Mechanics, find the equations $F_c = ma_c$ and $a_c = v^2/r$. Substitution of the second equation into the first gives $F_c = mv^2/r$. Substitute into this equation and solve for v.

52. $F_s = kx$

    k $\dfrac{F_s}{x}$

    $k = \dfrac{10.\,\text{N}}{0.25\,\text{m}}$

    $k = 40.\,\text{N/m}$

Explanation: Under Mechanics in the RT, find the equation $F_s = kx$. Substitute into the equation and solve for k.

53. E $\dfrac{F_e}{q}$

    E $\dfrac{3.60 \times 10^{-15}\,\text{N}}{1.60 \times 10^{-19}\,\text{C}}$

    E $2.25 \times 10^4\,\text{N/C}$

Explanation: In the RT under Electricity, find the equation $E = F_e/q$. The charge q is the charge on the electron which is the elementary charge (e), found on the List of Physical Constants. Substitute into the equation and solve for E.

54.

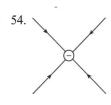

Explanation: The electric field lines are perpendicular to the surface of the charged sphere. By definition, the direction of an electric field, shown by the electric field lines, is from positive to negative. Therefore, the lines are directed toward the negatively charged sphere.

55.

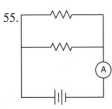

Explanation: Under Circuit Symbols in the RT, find the correct symbols for a battery, resistor and ammeter. To be in parallel with each other, the two resistors must provide separate paths for current. To measure the total circuit current, the ammeter must be in series with the parallel combination of resistors.

56. P $\dfrac{V}{R}$

    R $\dfrac{V}{P}$

    R $\dfrac{(120\,\text{V})}{900\,\text{W}}$

    R $16\,\Omega$

Explanation: Under Electricity in the RT, find the equation $P = V^2/R$. Substitute into the equation and solve for R.

57. — north and south   — perpendicular to spring
      — up and down   — left and right

Explanation: In a transverse wave, the particle motion is perpendicular to the direction of travel of the wave through the slinky.

58. 1.5 m

Explanation: The wavelength of the wave is the distance between two successive compressions.

59. The wavelength is shorter when the speaker is moving.

Explanation: This is an example of the Doppler effect, caused by relative motion between the source of a wave and a receiver. When the source is moving toward the observer, the wavelength is shorter than when the source is at rest.

60. 16 m/s

Explanation: In the RT under Mechanics, find the equation $\bar{v} = d/t$. The horizontal motion is with constant velocity. In the horizontal, the car travels a distance of 16.0 m in 1.00 s to reach point B. Substitution and solving gives $\bar{v} = (16.0 \text{ m})/(1.00 \text{ s}) = 16$ m/s.

61. 4.9 m/s

Explanation: In the RT under Mechanics, find the equation $v_f = v_i + at$. In the vertical, the car undergoes an acceleration equal to that of gravity (g), which is found on the List of Physical Constants. The car starts from rest in the vertical ($v_i = 0$). Substitution into the equation gives $v_f = 0 + (9.81 \text{ m/s}^2)(0.50 \text{ s})$. Solving, $v_f = 4.9$ m/s.

62. $d \quad vt_i \quad a\frac{1}{2} \quad {}^2$

$d_y \quad (4.9 \text{ m/s}) (0.50 \text{ s}) + \frac{1}{2}(9.81 \text{ m/s}^2) (0.50 \text{ s})^2$

$d_y \quad 3.7 \text{ m}$

Explanation: Under Mechanics in the RT, find the equation $d = v_i t + \frac{1}{2} at^2$. At point A, the vertical velocity of the car is 4.9 m/s (from question 61). The acceleration is that due to gravity. The time needed to travel from point A to point B is 0.50 s (1.00 s - 0.50 s). Substitute into the equation and solve for d.

63. $\Delta KE = \Delta PE = mg\Delta h$

$$\Delta h = \frac{\Delta KE}{mg}$$

$$\Delta h = \frac{3.13 \times 10^5}{(290. \text{ kg})(9.81 \text{ m/s}^2)}$$

$$\Delta h = 110. \text{ m}$$

Explanation: In the RT under Mechanics, find the equation $\Delta PE = mg\Delta h$. Neglecting friction, the kinetic energy of the car at the bottom of the hill will be equal to the potential energy of the car at the top of the hill. Therefore, $\Delta PE = \Delta KE = 3.13 \times 10^5$ J. The value of g is found on the List of Physical Constants in the RT. Substitute into the equation and solve for Δh.

64. $KE = \frac{1}{2} mv^2$

$$v = \sqrt{\frac{2KE}{m}}$$

$$v = \sqrt{\frac{2(3.13 \times 10^5)}{290. \text{ kg}}}$$

$$v = 46.5 \text{ m/s}$$

Explanation: Under Mechanics in the RT, find the equation $KE = \frac{1}{2}mv^2$. The kinetic energy of the car at the bottom of the hill is given. Substitute into the equation and solve for v.

65. $a = \frac{\Delta v}{t}$

$$a = \frac{46.5 \text{ m/s}}{5.3 \text{ s}}$$

$$a = 8.8 \text{ m/s}^2$$

Explanation: Under Mechanics in the RT, find the equation $a = \Delta v/t$. Using the speed of the car calculated in question 64 and knowing the car started from rest at the top of the hill, $\Delta v = 46.5 \text{ m/s} - 0 = 46.5 \text{ m/s}$. Substitute into the equation and solve for a.

66. 3.0 m *or* 3 m

Explanation: One wavelength of a wave consists of one crest and one trough. This is what is shown in the diagram. Therefore, the wavelength of the standing wave is 3.0 m.

67. $v = f\lambda$

$v = (20.0 \text{ Hz})(3.0 \text{ m})$

$v = 60. \text{ m/s}$

Explanation: Under Waves in the RT, find the equation $v = f\lambda$. Using the wavelength found in question 66 and the frequency given for the wave, substitute into the equation and solve for v.

68. $n_1 \sin \theta_1 = n_2 \sin \theta_2$

$$\sin \theta_2 = \frac{n_1 \sin \theta_1}{n_2}$$

$$\sin \theta_2 = \frac{(1.00)(\sin 30.°)}{1.50}$$

$$\theta_2 = 19°$$

Explanation: In the RT under Waves, find the equation $n_1 \sin \theta_1 = n_2 \sin \theta_2$. n_1 and n_2 are the absolute indices of refraction of air and lucite, respectively, and are found in the RT on the table of Absolute Indices of Refraction. Substitute into the equation and solve for θ_2.

69.

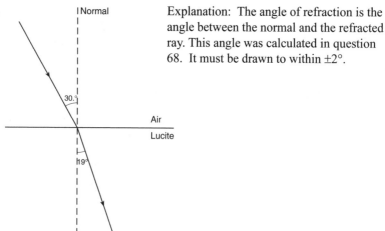

Explanation: The angle of refraction is the angle between the normal and the refracted ray. This angle was calculated in question 68. It must be drawn to within ±2°.

70. green

Explanation: Find the Electromagnetic Spectrum chart in the RT. In the visible light portion, a frequency of 5.48×10^{14} Hz places this photon in the green portion of the visible spectrum.

71. $E_{photon} \quad hf$

$E_{photon} \quad (6.63 \times 10^{-34} \cdot s)(5.48 \quad 10 \quad Hz$

$E_{photon} \quad 3.63 \times 10^{-19}$

Explanation: Under Modern Physics in the RT, find the equation $E_{photon} = hf$. The value of h is found on the List of Physical Constants in the RT. Substitute into the equation and solve for E_{photon}.

72. 2.27 eV

Explanation: On the List of Physical Constants in the RT, $1 \text{ eV} = 1.60 \times 10^{-19}$ J. Use this as a conversion factor to convert the answer in J from question 71 to eV: 3.63×10^{-19} J $\times \dfrac{(1 \text{ eV})}{(1.60 \times 10^{-19} \text{ J})} = 2.27$ eV.

Part A

1. 2   Distance is a scalar quantity, having magnitude or size only. Displacement is a vector quantity, having both magnitude and direction. Distance is the magnitude of the displacement vector. The total distance traveled by the player is 33.4 meters (27.4 m + 3.0 m + 3.0 m). The magnitude of the displacement of the player from the batter's box is 27.4 meters, the distance from the batter's box to first base.

2. 3   Velocity is a vector quantity, having both magnitude and direction. Speed is the magnitude of velocity vector. The velocity of 5.0 m/s east and that of 3.3 m/s south are at a right angle to each other. The resultant velocity of the boat is the hypotenuse of the right triangle formed by these two velocities. The magnitude of this resultant velocity may be calculated using Pythagorean Theorem $(c^2 = a^2 + b^2)$: $c^2 = (5.0 \text{ m/s})^2 + (3.3 \text{ m/s})^2$. Solving, $c = 6.0$ m/s. See Geometry and Trigonometry - Right Triangle in the RT.

3. 3   Under Mechanics in the RT, find the equation $\bar{v} = \dfrac{d}{t}$. The average speed can be calculated using

$$\bar{v} = \frac{(v_i + v_f)}{2} = \frac{(15.0 \text{ m/s} + 21.0 \text{ m/s})}{2} = 18.0 \text{ m/s} \cdot \text{ Substitution}$$

into the first equation gives $18.0 \text{ m/s} = \dfrac{d}{(12.0 \text{ s})}$. Solving, $d = 216$ m.

4. 4   In the RT under Mechanics, find the equations $J = F_{net}t = \Delta p$ and $p = mv$. The change in momentum (Δp) can be expressed as $m\Delta v$, where Δv is the change in velocity of the baseball. Therefore $F_{net}t = m\Delta v$. Since the baseball is brought to rest from 15 m/s, $\Delta v = 15$ m/s. Substitution gives $F_{net}(0.040 \text{ s}) = (0.149 \text{ kg})(15.0 \text{ m/s})$. Solving, $F_{net} = 56$ N.

5. 1   The vertical and horizontal motion of the stone are independent of each other. Therefore, the time for the stone to reach the ground is independent of the horizontal speed. Under Mechanics in the RT, find the equation $d = v_it + \frac{1}{2}at^2$. In the vertical, $v_i = 0$ and $a = g$, the acceleration due to gravity, found on the List of Physical Constants in the RT. Substitution gives $50.0 \text{ m} = 0 + \frac{1}{2}(9.81 \text{ m/s}^2)(t^2)$. Solving, $t = 3.19$ s.

6. 4   An object in equilibrium is at rest or moving with a constant velocity.

7. 2   The mass of an object is constant. Therefore, the mass of the spacecraft a distance of one Earth radius above the Earth's surface is the same as its mass at the Earth's surface.

8. 1 Newton's Third Law states that for every action, there is an equal and opposite reaction. The magnitude of the force that the sled exerts on the student equals the magnitude of the force that the student exerts on the sled.

9. 4 Inertia is measured quantitatively by an objects mass. The greater the mass, the greater the inertia.

10. 4 In the RT under Mechanics, find the equation $a = \dfrac{F_{net}}{m}$. Since the blocks move as a unit, the total mass being accelerated is 1.0 kg + 2.0 kg = 3.0 kg. Substitution and solving gives
$a = \dfrac{(12\ \text{N})}{(3.0\ \text{kg})} = 4.0\ \text{m}$.

11. 3 Under Mechanics in the RT, find the equation $v_f^2 = v_i^2 + 2ad$. At the maximum height, the velocity of the ball is 0 m/s ($v_f = 0$). The acceleration a is that due to gravity, g, which is found on the List of Physical Constants in the RT. Substitution gives
$0 = (29.4\ \text{m/s})^2 + 2(9.81\ \text{m/s}^2)(d)$. Solving, $d = 44.1$ m.

12. 3 The term centripetal means pointing toward the center or center seeking. Therefore, the centripetal force acting of the mass is directed toward point C.

13. 2 In the RT under Mechanics, find the equations $p_{before} = p_{after}$ and $p = mv$. Before firing, the bullet and the gun are at rest, therefore $p_{before} = 0$. Substitution gives $0 = (3.1\ \text{kg})(v) + (0.015\ \text{kg})(500.\ \text{m/s})$. Solving, $v = -2.4$ m/s. The negative sign for the velocity indicates that the gun moves in the opposite direction of the bullet.

14. 4 The horizontal and vertical motions are independent of each other. The horizontal distance traveled depends only upon the initial horizontal speed and the time of flight. In the RT under Mechanics, find the equation $\bar{v} = \dfrac{d}{t}$. Since the horizontal speeds are constant, they are the average speeds. Substitution of the initial horizontal speed and time of flight into the equation and solving for d gives 240. m, 240. m, 250. m, and 320. m for the horizontal distances traveled for projectiles A, B, C and D, respectively.

15. 2 In the RT under Mechanics, find the equation $\Delta PE = mg\Delta h$. Also under Mechanics, find the equation $g = F_g/m$. Solving for F_g, $F_g = mg$, which is the weight of the box (155 N). The first equation may now be written $\Delta PE = F_g\Delta h$. Substitution and solving gives
$\Delta PE = (155\ \text{N})(1.80\ \text{m}) = 279$ J.

16. 3   In the RT under Mechanics, find the equation $E_T = PE + KE + Q$. For the car, PE is the elastic potential energy of the wound spring. Since the car coasts to a stop, the elastic potential energy and kinetic energy are both zero. Therefore, the elastic potential energy and kinetic energy have been converted to internal energy or heat energy.

17. 3   Under Mechanics in the RT, find the equation $KE = \frac{1}{2}mv^2$.
Substitution and solving gives $KE = \frac{1}{2}(75 \text{ kg})(12 \text{ m/s})^2 = 5.4 \times 10^3 \text{ J}$.

18. 2   Under Electricity in the RT, find the equation $W = Pt = VIt = I^2Rt = V^2t/R$. Since potential difference, current, resistance and time are given, you can use any of these to solve for W. For example,
$W = VIt = (120. \text{ V})(8.00 \text{ A})(60.0 \text{ s}) = 5.76 \times 10^4 \text{ J}$.

19. 1   Magnetic fields are produce by moving charged particles.

20. 2   In the RT under Electricity, find the equation $I = \Delta q/t$. Substitution and solving gives $I = (30. \text{ C})/(6.0 \text{ s}) = 5.0 \text{ A}$.

21. 3   Under Electricity in the RT, find the equation $F_e = \dfrac{kq_1q_2}{r^2}$. q_1 and q_2 are the charges on the electrons (the elementary charge), which along with k, are found on the List of Physical Constants on the RT. Substitution gives $F_e = \dfrac{\left(8.99 \times 10^9 \text{ N}\bullet\text{m}^2/\text{C}^2\right)\left(1.60 \times 10^{-19} \text{ C}\right)\left(1.60 \times 10^{-19} \text{ C}\right)}{\left(1.00 \times 10^{-8} \text{ m}\right)^2}$.
Solving, $F_e = 2.30 \times 10^{-12} \text{ N}$.

22. 2   By definition, the direction of an electric field is from positive to negative. Therefore, sphere B must be positive and sphere A must be negative.

23. 3   In choices 1 and 3, the resistances are connected in parallel. Under Electricity-Parallel Circuits, find the equation $1/R_{eq} = 1/R_1 + 1/R_2 + 1/R_3 + ...$ Substitution into this equation and solving for R_{eq} gives 1 Ω and 0.7 Ω for 1 and 3, respectively. In choices 2 and 4, the resistances are connected in series. Under Electricity - Series Circuits, find the equation $R_{eq} = R_1 + R_2 + R_3 + ...$. Substitution into this equation and solving for R_{eq} gives 4 Ω and 8 Ω for 2 and 4, respectively.

24. 4   In a longitudinal wave, the direction of particle motion is parallel to the direction of wave movement.

25. 2   The wavelength of a wave is defined as the distance between two consecutive particles on a wave that are in phase. To be in phase, the particles have the same displacement from the rest or equilibrium position and are moving in the same direction.

26. 1   The amplitude of a wave is a measure of the energy carried by the wave.

27. 4   For no refraction or bending of the wave to occur as it passes from flint glass into medium X, the absolute index of refraction of medium X must be equal to that of flint glass. Referring to the Absolute Indices of Refraction table in the RT, the index of refraction of flint glass is 1.66.

28. 3   Under Waves in the RT, find the equation $n = c/v$. The value of c, the speed of light in a vacuum, is on the List of Physical Constants in the RT. Substitution and solving gives $n = \dfrac{\left(3.00 \times 10^8 \text{ m/s}\right)}{\left(1.71 \times 10^8 \text{ m/s}\right)} = 1.75$.

29. 1   Resonance is defined as the response of an object to impressed vibrations of the same frequency as the natural frequency of vibration of the object.

30. 1   All electromagnetic waves travel at the same speed through a vacuum, known as the speed of light in a vacuum (c).

31. 3   In the RT, find the Particles of the Standard Model table. The charge of an up quark is $+\frac{2}{3}$ e and that of a down quark is $-\frac{1}{3}$ e. The charge on a particle composed of two up quarks and one down quark is $\left(+\frac{2}{3}e\right) + \left(+\frac{2}{3}e\right) + \left(-\frac{1}{3}e\right) = +1$ e. This is the charge on a proton. A positron, also with a charge of +1 e, is the antiparticle of the electron. These are not composed of quarks.

32. 4   The strong force is the force that holds the nucleons (protons and neutrons) together the nucleus of an atom.

33. 1   Under Modern Physics in the RT, find the equation $E = mc^2$. The value of c (speed of light in a vacuum) is found on the List of Physical Constants in the RT. Substitution gives 1.03×10^{-13} J $= m(3.00 \times 10^8$ m/s$)^2$. Solving, $m = 1.14 \times 10^{-30}$ kg.

34. 3   An antimatter particle has the same mass but opposite charge of the corresponding matter particle..

35. 2   This is caused by the Doppler Effect, which is the change in the observed frequency of a wave due to relative motion between the observer and the source of the wave. When the source of a wave is moving away from the observer, the observed frequency is lower than the actual frequency of the source.

Part B-1

36. **2** To answer this question, think in terms of your familiarity with the standard of mass in the Metric System, the kilogram (kg). In the RT under Mechanics, find the equation $\Delta PE = mg\Delta h$. The change in potential energy of the physics textbook as it is lifted is equal to the work done in lifting it the 0.10 m. The value of g (acceleration due to gravity) is on the List of Physical Constants in the RT. Use the equation for ΔPE to solve for the mass of the textbook in each of the four choices. For choices 1 through 4, they are, respectively, 0.15 kg, 1.5 kg, 15 kg and 150 kg. Choice 2 (1.5 kg) is the only reasonable mass of a typical physics textbook. Also, 1 kg = 2.2 lb in our system.

37. **4** The joule is a unit of work. Under Electricity in the RT, find the equation $V = W/q$. Solving for W, $W = Vq$. The unit for potential difference is the volt. The unit of charge is the coulomb. The joule is therefore the product of the units of potential difference and charge.

38. **2** In the RT under Mechanics, find the equations $P = Fd/t$ and $g = F_g/m$. The force needed to lift an object at constant speed is equal to the weight of the object. Solving the second equation for F_g, $F_g = mg$. The first equation may now be written as $P = mgd/t$. The value of g is found on the List of Physical Constants in the RT. Substitution and solving gives $P = \dfrac{(0.50 \text{ kg})(9.81 \text{ m/s}^2)(1.5 \text{ m})}{(5.0 \text{ s})} = 1.5$ W.

39. **4** Mechanical energy is the total kinetic and potential energy of an object. In the absence of friction, the change in kinetic energy equals the change in potential energy. Therefore, the mechanical energy of an object remains constant.

40. **3** In the absence of friction, the kinetic energy of the child at the bottom of the slide equals the potential energy of the child at the top of the slide. Under Mechanics in the RT, find the equations $KE = \frac{1}{2}mv^2$ and $\Delta PE = mg\Delta h$. Equating the KE and ΔPE gives $\frac{1}{2}mv^2 = mg\Delta h$. In this equation, the mass of the child cancels out, leaving $\frac{1}{2}v^2 = g\Delta h$. The value of g is found on the List of Physical Constants in the RT. Substitution into the equation gives $\frac{1}{2}(7.0 \text{ m/s})^2 = (9.81 \text{ m/s}^2)(\Delta h)$. Solving, $\Delta h = 2.5$ m.

41. **2** Under Electricity in the RT, find the equation $R = V/I$. Choose any point that lies on the line representing the relationship between current and potential difference and substitute into this equation. For example, when the current is 0.5 A, the corresponding potential difference is 1.0 V. Substitution and solving gives $R = \dfrac{(1.0 \text{ V})}{(0.50 \text{ V})} = 2.0\,\Omega$.

June 2010 Answer Key

42. 4   Velocity is a vector quantity, having both magnitude and direction. On the upward trip, the velocity of the baseball decreases uniformly to 0. On the downward trip, the velocity of the baseball increases uniformly from 0 to a value equal to that of the initial upward velocity. If vertically upward is considered to be the positive direction, then vertically downward must be considered the negative direction. Graph 4 illustrates the upward and downward trip for the baseball.

43. 3   Under Mechanics in the RT, find the equation $d = v_i t + \frac{1}{2} at^2$. Since the object starts from rest, $v_i = 0$. Substitution into the equation gives 22 m = $0 + \frac{1}{2}(a)(3.0 \text{ s})^2$. Solving, $a = 4.8$ m/s^2. This is the acceleration due to gravity near the surface of the planet. The acceleration due to gravity near the Earth's surface (g, found on the List of Physical Constants in the RT) is 9.81 m/s^2. The acceleration due to gravity near the surface of the planet is approximately one-half of this value.

44. 3   The work done to increase the speed of the object is equal to the increase in kinetic energy of the object. In the RT under Mechanics, find the equation $KE = \frac{1}{2} mv^2$. The initial KE of the object is $KE = \frac{1}{2}(15.0 \text{ kg})(7.50 \text{ m/s})^2 = 422$ J. The final KE of the object is $KE = \frac{1}{2}(15.0 \text{ kg})(11.5 \text{ m/s})^2 = 992$ J. The change in KE is 570. J, which is the work done on the object.

45. 1   The resistances are connected in series. Under Electricity - Series Circuits in the RT, find the equation $R_{eq} = R_1 + R_2 + R_3 + ...$. Substitution and solving gives $R_{eq} = (4.0 \ \Omega) + (6.0 \ \Omega) + (8.0 \ \Omega) + (6.0 \ \Omega) = 24 \ \Omega$. Now under Electricity, find the equation $R = V/I$. Substitution gives $24 \ \Omega = (12 \text{ V})/I$. Solving, $I = 0.50$ A.

46. 4   Under Electricity in the RT, find the equation $P = V^2/R$. To obey Ohm's Law, we assume that the resistance remains constant. This equation then indicates the power expended by the resistor varies directly with the square of the potential difference applied to the resistor. This relationship is shown by graph 4. See 4b under Helpful Hints for Physics in the back of this book.

47. 1   In the RT under Mechanics, find the equation $F_g = Gm_1m_2/r^2$. Under Electricity in the RT, find the equation $F_e = kq_1q_2/r^2$. In both of these equations, the force varies inversely with the square of the distance between the particles. The graphs in choice 1 show this relationship. See 4d under Helpful Hints for Physics in the back of this book.

48. 3 Under Waves in the RT, find the equation $v = f\lambda$. The distance shown in the diagram represents 1.5 wavelengths. Therefore, the wavelength of the wave is $\dfrac{(6.0 \text{ m})}{(1.5)} = 4.0 \text{ m}$. Substitution into the equation gives $v = (2.0 \text{ Hz})(4.0 \text{ m})$. Solving, $v = 8.0 \text{ m/s}$.

49. 1 The angle of incidence is the angle between the incident ray and normal at the point of incidence. The angle of incidence is then $90° - 65° = 25°$. The angle of reflection is the angle between the reflected ray and the normal, which is equal to the angle of incidence.

50. 3 A node on a standing wave is a point of minimum or no vibration. An antinode on a standing wave is a point of maximum vibration.

Part B-2

51. 25 m/s ± 1 m/s

Explanation: Find 40. m/s on the Initial Velocity axis and read straight up until that line intersects the graph. Now read over to the Initial Vertical Velocity axis and the value is 25 m/s.

52. 39° ± 2°

Explanation: The initial velocity is the hypotenuse of the right triangle formed by the initial vertical velocity and the initial horizontal velocity. Under Geometry and Trigonometry - Right Triangle in the RT, find the equation $\sin \theta = a/c$. a is magnitude of the initial vertical velocity (v_{iy}) solved for in question 51 and c is the magnitude of the initial velocity (v_i) given in question 51. Substitution gives $\sin \theta = \dfrac{(25 \text{ m/s})}{(40. \text{ m/s})}$. Solving, $\sin \theta = 0.63$ and $\theta = 39°$.

53. $\begin{aligned} v_{ix} &= v_i \cos \theta \\ v_{ix} &= (40. \text{ m/s})\cos 39° \\ v_{ix} &= 31 \text{ m/s} \end{aligned}$ *or* $\begin{aligned} v_{ix}^{2} + v_{iy}^{2} &= v_i^{2} \\ v_{ix} &= \sqrt{v_i^{2} - v_{iy}^{2}} \\ v_{ix} &= \sqrt{(40. \text{ m/s})^2 - (25 \text{ m/s})^2} \\ v_{ix} &= 31 \text{ m/s} \end{aligned}$

Explanation: Under Mechanics in the RT find the equation $A_x = A \cos \theta$. A_x is the initial horizontal velocity and A is the initial velocity.

or

In the RT under Geometry and Trigonometry - Right Triangle, find the equation $c^2 = a^2 + b^2$ (the Pythagorean Theorem). Side a is the initial vertical velocity, side b is the initial horizontal velocity and side c is the initial velocity.

54. — friction
— Some of the gravitational energy of the mass was converted into internal energy. Therefore, it could not return to its original height.
— air resistance

Explanation: The initial gravitational potential energy of the pendulum mass has been lost due to friction, air resistance or has be changed into internal energy.

55.

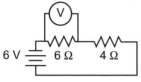

Explanation: The resistances provide a single path for current when connected in series. The voltmeter must be connected in parallel with the device that it measures the potential difference across.

56. $PE_s = \frac{1}{2}kx^2$

$k = \frac{2PE_s}{x^2}$

$k = \frac{2(1.25 \times 10^{-2}\ \text{J})}{(2.50 \times 10^{-2}\ \text{m})^2}$

$k = 40.0\ \text{N/m}$

Explanation: Under Mechanics in the RT, find the equation $PE_s = \frac{1}{2}kx^2$. Using the given values of the change in length of the spring and the total potential energy stored in the spring, substitute into the equation and solve for k.

57. $6.25 \times 10^{-2}\ \Omega$

Explanation: In the RT under Electricity, find the equation $R = V/I$. Substitution and solving gives $R = \frac{(1.5\ V)}{(24.0\ A)} = 6.25 \times 10^{-2}\ \Omega$.

58. $R = \frac{\rho L}{A}$

$\rho = \frac{RA}{L}$

$\rho = \frac{(6.25 \times 10^{-2}\ \Omega)(3.14 \times 10^{-6}\ \text{m}^2)}{3.50\,\text{m}}$

$\rho = 5.61 \times 10^{-8}\ \Omega \bullet \text{m}$

Explanation: Under Electricity in the RT, find the equation $R = \rho L/A$. Using the value of R from question 57 and the given values of L and A, calculate the resistivity.

59. 6.3 m/s

Explanation: Under Mechanics in the RT, find the equation $\bar{v} = \frac{d}{t}$. The distance traveled in 10. s is 10 times the circumference of the circle (10 C). In the RT under Geometry and Trigonometry - Circle, find the equation $C = 2\pi r$. The equation may now be written as $\bar{v} = \frac{10(2\pi r)}{t}$. Substitution gives $\bar{v} = \frac{10(2 \times 3.14 \times 1.0\ \text{m})}{1.0\ \text{s}}$. Solving, $v = 6.3$ m/s.

60. $F_c = ma_c \qquad a_c = \dfrac{v^2}{r}$

$F_c = \dfrac{mv^2}{r}$

$F_c = \dfrac{(0.028 \text{ kg})(6.3 \text{ m/s})^2}{1.0 \text{ m}}$

$F_c = 1.1 \text{ N}$

Explanation: Under Mechanics in the RT, find the equations $F_c = ma_c$ and $a_c = v^2/r$. Substituting the second equation into the first, it becomes $F_c = mv^2/r$. Using the speed of the object from question 59 and the given values of m and r, substitute into the equation and calculate F_c.

61-63.

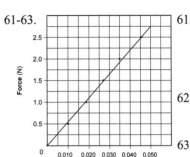

Elongation (m)

61. Explanation: Make sure that your scale is linear (each division shows equal increments) and uses most of the vertical axis.

62. Plot the points accurately to ± 0.3 grid spaces. Refer to the graph in question 61.

63. Draw the best-fit line through the plotted points. Refer to the graph in question 61.

64. $k = \dfrac{\Delta F}{\Delta x}$

$k = \dfrac{2.5 \text{ N}}{0.046 \text{ m}}$

$k = 54 \text{ N/m}$

or

$\text{slope} = \dfrac{\Delta y}{\Delta x}$

$\text{slope} = \dfrac{2.5 \text{ N} - 0.8 \text{ N}}{0.046 \text{ m} - 0.015 \text{ m}}$

$\text{slope} = 55 \text{ N/m}$

Explanation: Under Mechanics in the RT, find the equation $F_x = kx$. Solving for k, $k = F_x/x$. F_x may represent the change in force and x may represent the change in length of the spring. Using the data from the Data Table, substitute the values into the equation and calculate k.

or

Using the definition of slope as the change in y divided by the change in x, use values from the best fit line to calculate the slope of the graph which represents k. See 4e under Helpful Hints for Physics in the back of this book.

65. $a \quad \dfrac{F_{net}}{m}$

$F_{net} \quad ma$

$F_{net} \quad (20. \text{ kg})(1.4 \text{ m/s}^2)$

$F_{net} \quad 28 \text{ N}$

Explanation: In the RT under Mechanics, find the equation $a = F_{net}/m$. Substitute into the equation the given values of m and a and solve for F_{net}.

June 2010
Answer Key

66.

Level ice
(Not to Scale)

Explanation: Using the scale of 1.0 centimeter = 5.0 newtons, the vector representing a force of 28 N (answer to question 65) must be 5.6 cm long originating from point A and pointing to the right.

67. 2.0×10^2 N **or** 196 N

Explanation: The normal force acting upward on the block must be equal to the weight of the block, which acts downward. Under Mechanics in the RT, find the equation $g = F_g/m$. Solving for the weight of the block, $F_g = mg$. The value of g is found on the List of Physical Constants in the RT. Substitution gives $F_g = (20.\text{ kg})(9.81 \text{ m/s}^2)$. Solving, $F_g = 196$ N.

68. $F_f \quad F_N$

$F_f \quad (0.28)(2.0 \times 10^2 \text{ N})$

$F_f \quad 56$ N

Explanation: Under Mechanics in the RT, find the equation $F_f = \mu F_N$. The value of F_N is the weight of the block calculated in question 67. Using the value of 2.0×10^2 N and the given coefficient of friction, calculate the magnitude of the force of friction acting on the block.

69. $50.° \pm 2°$

Explanation: The angle of incidence is the angle between the incident ray and the normal at the point of incidence. Using a protractor, measure this angle.

70. $n_1 \sin \theta_1 = n_2 \sin \theta_2$

$\sin \theta_2 = \dfrac{n_1 \sin \theta_1}{n_2}$

$\sin \theta_2 = \dfrac{1.00 (\sin 50.°)}{1.50}$

$\theta_2 = 31°$

Explanation: Under Waves in the RT, find the equation $n_1 \sin \theta_1 = n_2 \sin \theta_2$. n_1 an n_2 are the absolute indices of refraction of the mediums in which the angle of incidence and angle of refraction are measured, respectively. These are found in the RT on the Table of Absolute Indices of Refraction. θ_1 is the angle of incidence measured in question 69. Substitute the given values into the equation and solve for θ_2.

71. $50°$

Explanation: Since the top and bottom surfaces of the Lucite are parallel to each other, the ray emerging from the Lucite into air must be parallel to the original incident ray.

72. 1.24 eV

Explanation: Under Modern Physics in the RT, find the equation
$E_{photon} = E_i - E_f$. The energy of the levels of the mercury atom are found on
the Energy Level Diagrams in the RT. Substitution into the equation gives
$E_{photon} = (-4.95 \text{ eV}) - (-3.71 \text{ eV})$. Solving, $E_{photon} = -1.24 \text{ eV}$. Be very
careful of the negative signs in the equation. The negative value for the
photon energy indicates that it is absorbed.

73. $1.98 \times 10^{-19} \text{ J}$

Explanation: The relationship between the electron volt and the joule is
found in the RT on the List of Physical Constants ($1 \text{ eV} = 1.60 \times 10^{-19} \text{ J}$).
Use this as a conversion factor to calculate the energy in joules:
$1.24 \text{ eV} \times (1.60 \times 10^{-19} \text{ J})/(1 \text{ eV}) = 1.98 \times 10^{-19} \text{ J}$.

74. $E_{photon} \quad hf$

$f \quad \dfrac{E_{photon}}{h}$

$f \quad \dfrac{1.98 \times 10^{-19}}{6.63 \times 10^{-34}} \quad \bullet \; \text{\textit{}}$

$f = 2.99 \times 10^{14} \text{ Hz}$

Explanation: Under Modern Physics in the RT, find
the equation $E_{photon} = hf$. The value of h (Planck's
Constant) is found on the List of Physical Constants
in the RT. Use the photon energy in J from
question 73 and solve for f.

75. infrared

Explanation: In the RT, find The Electromagnetic Spectrum table. The
frequency calculated in question 74 places this radiation in the infrared
family.

June 2011
Part A

1.   1   A scalar is a quantity with magnitude or size only. A vector is a quantity with both magnitude and direction. In choice 1, speed is a scalar and velocity is a vector.

2.   2   Under Mechanics in the RT. find the equation $v_f^2 = v_i^2 + 2ad$.
Since the car starts from rest, $v_i = 0$.
Substitution gives $(15 \text{ m/s})^2 = (0 \text{ m/s})^2 + 2(a)(100. \text{ m})$.
Solving, $a = 1.1 \text{ m/s}^2$.

3.   1   In the RT under Mechanics, find the equation $a = \Delta v/t$, where Δv may be expressed as $(v_f - v_i)$. Substitution and solving gives $a = (8.0 \text{ m/s} - 3.0 \text{ m/s})/(2.0 \text{ s}) = 2.5 \text{ m/s}^2$.

4.   4   The acceleration of the rock is that due to gravity (g: 9.81 m/s^2), which is constant near the Earth's surface. This acceleration causes the speed of the rock to increase.

5.   3   Under Mechanics in the RT, find the equation $v_f = v_i + at$. For the upward trip, v_f at the highest point is 0 and the acceleration is $-g$, since the object is slowing down. The value of g is found on the List of Physical Constants. Substitution gives $0 = 15.0 \text{ m/s} + (-9.81 \text{ m/s}^2)(t)$. Solving, $t = 1.53 \text{ s}$. The time to come down to the same height from which it was thrown is the same as that needed to reach the highest point. The total time in the air is therefore $2(1.53 \text{ s}) = 3.06 \text{ s}$.

6.   1   If the elevator is at rest, the force the student exerts on the floor of the elevator is equal to the students weight. If the elevator is accelerating upward, the force the student exerts on the floor of the elevator is greater than the students weight due to the accelerating force acting on the student. If the elevator is accelerating downward, the force the student exerts on the floor of the elevator is less than the students weight since some of the force of gravity is acting to provide the acceleration of the student.

7.   1   Under Mechanics in the RT, find the equations $F_c = ma_c$ and $a_c = v^2/r$. Substituting the second equation into the first equation gives $F_c = mv^2/r$. From this equation, centripetal force varies inversely with the radius of the path. Increasing the radius will decrease the centripetal force.

8.   4   The force of gravity or the weight of the shuttle acts toward the center of the Earth. Therefore, the weight would act as the centripetal force on the shuttle. Also, weight is the only choice given that is a force.

9. 3   In the RT under Mechanics, find the equation $W = Fd$. The work done on the box would be $W = (25 \text{ N})(30. \text{ m}) = 750 \text{ J}$. Under frictionless conditions, this would be equal to the gain in kinetic energy. Since the floor is horizontal, there would be no increase in the potential energy of the box. Now find the equation $E_T = PE + KE + Q$, where E_T is equivalent to the work done on the box. Substitution gives $750 \text{ J} = 0 + 300. \text{ J} + Q$. Solving, $Q = 450 \text{ J}$.

10. 4   Under Mechanics in the RT, find the equations $P = Fd/t$ and $g = F_g/m$. In lifting an object at a constant speed, the force required is equal to the objects weight. Solving the second equation for F_g, $F_g = mg$. The first equation may now be written as $P = mgd/t$. The value of g found on the List of Physical Constants. Substitution and solving gives $P = (2.0 \text{ kg})(9.81 \text{ m/s}^2)(15 \text{ m})/(6.0 \text{ s}) = 49 \text{ W}$.

11. 2   Maximum horizontal distance or range of a projectile is achieved when the angle of projection is 45°.

12. 4   The net force acting on an object in equilibrium must be 0. As shown in the diagram, the net force acting on the object is 4 N toward the right. An additional force of 4 N toward the left is therefore needed to produce equilibrium.

13. 2   Inertia is a quantitative measure of the mass of an object. The greater the mass, the greater the inertia.

14. 3   Mechanical energy is the total kinetic and potential energy of an object. When pulled at a constant speed, the kinetic energy of the block remains the same. As it is pulled up an incline, the potential energy of the block increases. Therefore, the total mechanical energy increases.

15. 2   By definition, electric field lines are directed from positive to negative. The lines therefore point away from a positive charge and toward a negative charge.

16. 3   In reflection, the angle of incidence (angle between the incident ray and the normal) equals the angle of reflection (angle between the reflected ray and the normal). In the diagram, angle B is the angle of incidence and angle C is the angle of reflection.

17. 1   The force exerted on sphere A by sphere B must be equal to the force exerted on sphere B by sphere A. This is an example of Newton's Third Law, which states that for every action, there is an equal and opposite reaction.

18. 4   The positively charged particle will be repelled by the top plate and attracted by the bottom plate. Remember, like charges repel and unlike charges attract.

19. 2   Under Electricity in the RT, find the equation $V = W/q$. For a proton, q is equal to the elementary charge (e), which is found on the List of Physical Constants. Substitution gives 100. $V = W/(1.60 \times 10^{-19}$ C). Solving, $W = 1.60 \times 10^{-17}$ J.

20. 2   In the RT under Electricity, find the equation $I = \Delta q/t$. The time must be in seconds (2 min = 120 s). Substitution and solving gives $I = (240$ C)$/(120$ s) = 2.0 A.

21. 1   Under Electricity in the RT, find the equation $R = V/I$. Solving for I, $I = V/R$. If the potential difference is constant, the current varies inversely with the resistance. Doubling the resistance would halve the current. See 4c under Helpful Hints for Physics in this book.

22. 3   In any electrical circuit, the total potential drop must be equal to the voltage of the source, in this case, the battery. This is an example of the Law of Conservation of Energy.

23. 3   In the RT under Electricity, find the equation $W = I^2Rt$. Substitution and solving gives $W = (0.50$ A$)^2(4.0 \ \Omega)(10.$ s) = 10. J.

24. 1   If the wire cuts across the magnetic field lines, a potential difference is produced across the conductor. This is electromagnetic induction or the generator effect.

25. 1   A wave is simply a means of transferring energy from one point to another. Mass does not travel along with a wave.

26. 4   The amplitude of a wave is the maximum displacement of the wave from the rest or equilibrium position. The equilibrium position is represented by the solid horizontal line (0 cm displacement) in the diagram.

27. 3   The period of a wave is the time needed to generate one wave or pulse. For this wave, the period is (10. s)/(4.0 waves) = 2.5 s.

28. 2   For two points on a wave to be 90° out of phase with each other, they must be separated by a distance equal to one-quarter wavelength. One wavelength of this wave consists of one crest and one trough.

29. 2   Under Waves in the RT, find the equation $v = f\lambda$. The speed of sound in air at STP is found on the List of Physical Constants. Substitution gives 3.31×10^2 m/s = (256 Hz)(λ). Solving, $\lambda = 1.29$ m.

30. 1   In the RT under Modern Physics, find the equation $E = mc^2$. The mass of the positron is equal to the mass of the electron (m_e), which is found on the List of Physical Constants. The total mass annihilated, that of the electron and positron, is then $2(9.11 \times 10^{-31}$ kg). The speed of light (c) is also found on the List of Physical Constants. Substitution and solving gives $E = 2(9.11 \times 10^{-31}$ kg)$(3.00 \times 10^8$ m/s)$^2 = 1.64 \times 10^{-31}$ J.

31. 3   A sound wave is a longitudinal wave and requires a medium for transmission. The amplitude of a wave is a measure of the energy carried by the wave.

32. 2   Resonance is the response of an object to a frequency that is the same as the natural frequency of vibration of the object.

33. 3   The Doppler Effect is the change in the frequency and wavelength of a wave caused by relative motion between the source and receiver of the wave. As the receiver (the spacecraft) approaches the source of the wave or signal (the Earth), the signal received has a higher frequency and shorter wavelength than the source.

34. 4   On the atomic scale, energy exhibits particle characteristics and particles exhibit wave characteristics. When electrons pass through a narrow slit, they show a diffraction pattern, acting as a wave. A photon is a particle of electromagnetic energy.

35. 4   Hadrons are baryons, which include protons and neutrons, which are found in the nucleus of atoms. The strong force is the force that holds the nuclear particles together. Electrons, found outside of the nucleus, are not affected by the strong force.

Part B-1

36. 2   Choices 1 through 4 are 2 cm, 20 cm, 200 cm, and 2000 cm, respectively. Choice 2 is the only reasonable choice. Remember that 1 in = 2.54 cm.

37. 1   Under Mechanics in the RT find the equations $a = \Delta v/t$ and $d = v_i t + \frac{1}{2} at^2$. From the graph, using the 0 to 3.0 s time interval, $a = (8.0$ m/s)/$(3.0$ s) = 2.7 m/s^2. Since the object starts from rest, $v_i = 0$. Substitution into the second equation and solving gives $d = 0 + \frac{1}{2} (2.7$ m/s$^2)(3.0$ s$)^2 = 12$ m.

38. 3   In the RT under Mechanics, find the equations $J = F_{net}t = \Delta p$ and $p = mv$. Since mass is a constant, a change in momentum (Δp) occurs due to a change in velocity (Δv). Therefore, $\Delta p = m\Delta v$. The first equation may now be written as $F_{net}t = m\Delta v$. Substitution gives $F_{net}(0.65 \text{ s}) = (75 \text{ kg})(6.0 \text{ m/s})$. Solving, $F_{net} = 692$ N.

39. 3   The force of friction is a horizontal force acting opposite to the direction of motion and since the wagon is moving at a constant velocity, it equals the horizontal component of the applied force. In the RT under Geometry and Trigonometry - Right Triangle, find the equation $\cos \theta = b/c$, where b is the horizontal component of the 22 N force. Substitution gives $\cos 35° = b/(22 \text{ N})$. Solving, $b = 18$ N. Therefore, the force of friction is 18 N.

40. 3   Under Electricity in the RT, find the equation $F_e = kq_1q_2/r^2$. This equation indicates that the electrical force varies directly with the charges and inversely with the square of the distance between their centers. The magnitude of the charge on sphere A is 3 times that on sphere B. Therefore, the force of sphere A on sphere C is 3 times as great as the force sphere B on sphere C. Sphere A is 2 times as far from sphere C as sphere B is from sphere C. Therefore, the force of sphere A on sphere C is ¼ as great as the force of sphere B on sphere C. The combined effect is (3)(¼) or ¾ as great. See 4a and 4d under Helpful Hints for Physics in this book.

41. 2   In the RT under Mechanics, find the equation $F_g = Gm_1m_2/r_2$. This equation indicates that the gravitational force varies inversely with the square of the distance between the centers. See 4d under Helpful Hints for Physics in this book.

42. 2   Under Mechanics in the RT, find the equation $\Delta PE = mg\Delta h$. This equation indicates that gravitational potential energy varies directly with the height above the surface of the Earth. See 4a under Helpful Hints for Physics in this book.

43. 4   Under Electricity in the RT, find the equation $E = F_e/q$. Solving for F_e, $F_e = Eq$. This shows that the electrostatic force varies directly with electric field strength. See 4a under Helpful Hints for Physics in this book.

44. 4   The two resistors are connected in series. Under Electricity - Series Circuits in the RT, find the equation $R_{eq} = R_1 + R_2 + R_3 + \dots$. Substitution and solving gives $R_{eq} = 10. \ \Omega + 20. \ \Omega = 30. \ \Omega$. Now find the equation $R = V/I$. Substitution gives $30. \ \Omega = (120 \ V)/(I)$. Solving, $I = 4.0$ A.

45. 1   By definition, the direction of magnetic field lines are from north to south. Therefore, A is a north pole and B is a south pole.

46. 2   In a transverse wave, the particle motion is perpendicular to the direction of wave travel. As the trough approaches, the cork will move down, then up as the crest approaches and then down as the crest passes.

47. 4   As the waves pass through the opening, they spread out to fill the region behind the opening, eventually reaching point P. This is known as diffraction.

48. 3   Nodes are points of no vibration on a standing wave. Antinodes are points of maximum vibration on a standing wave.

49. 3   On the RT, find the Classification Matter chart. Hadrons are baryons, which are composed of three quarks. The names and charges of the quarks are given in the RT on the Particles of the Standard Model chart. The charge on a proton is $+1$ e, where e is the elementary charge. It must be composed of 2 up quarks and 1 down quark, giving a charge of $2 \times (+2/3 \text{ e}) + (-1/3 \text{ e}) = +1$ e. The neutron is neutral. It must be composed of 1 up quark and 2 down quarks, giving a charge of $(+2/3 \text{ e}) + 2 \times (-1/3 \text{ e}) = 0$. Therefore, there are 3 up quarks and 3 down quarks in a deuterium nucleus.

50. 1   The amplitude of the resultant wave is the sum of the amplitudes of the two individual waves. The superposition produces the greatest positive displacement at point A: $(+0.10 \text{ m}) + (+0.20 \text{ m}) = +0.30 \text{ m}$.

Part B-2

51. $a = \dfrac{F_{net}}{m}$ Explanation: Under Mechanics in the RT,
find the equation $a = F_{net}/m$. Substitute the
$F_{net} = ma$ given values into the equation.

$F_{net} = (0.50 \text{ kg})(3.0 \text{ m/s}^2)$

52. $F_{net} = 1.5 \text{ N}$

Explanation: Solve the equation in question 51 for F_{net}.

53. 850 N

Explanation: On a horizontal surface, magnitude of the normal force exerted by the snow on the skis is equal to the weight of the student and the skis. This is an example of Newton's third Law.

54. $F_f = \mu F_N$

$F_f = (0.05)(850 \text{ N})$

Explanation: In the RT under Mechanics, find the equation $F_f = \mu F_N$. The value of μ is found on the table of Approximate Coefficients of Friction in the RT. Since the skier is moving, use the kinetic coefficient. Substitute the values into the equation.

55. $F_f = 40 \text{ N}$

Explanation: Solve the equation in question 54 for F_f.

56. $KE = \frac{1}{2}mv^2$

$KE = \frac{1}{2}(3.34 \times 10^{-27} \text{ kg})(2.89 \times 10^5 \text{ m/s})^2$

Explanation: In the RT under Mechanics, find the equation $KE = \frac{1}{2}mv^2$. Substitute the given values into the equation.

57. $KE = 1.39 \times 10^{-16} \text{ J}$

Explanation: Solve the equation in question 56 for KE.

58. $20. \ \Omega$

Explanation: The two resistors may be connected in series or parallel to the battery. Under Electricity - Series Circuits, find the equation $R_{eq} = R_1 + R_2 + R_3 + \dots$. Substitution and solving gives $100. \ \Omega + 25 \ \Omega = 125 \ \Omega$. Under Electricity - Parallel Circuits in the RT, find the equation $1/R_{eq} = 1/R_1 + 1/R_2 + 1/R_3 + \dots$. Substitution gives $1/R_{eq} = 1/(100. \ \Omega) + 1/(25 \ \Omega)$. Solving, $R_{eq} = 20. \ \Omega$. The smallest equivalent resistance is $20 \ \Omega$.

59. Power or the rate at which work is done

Explanation: The slope of the graph is work/time, which is the definition of power.

60. $P = \dfrac{V^2}{R}$

$P = \dfrac{(12 \text{ V})^2}{1.2 \ \Omega}$

Explanation: Under electricity in the RT, find the equation $P = V^2/R$. Substitute the given values into the equation.

61. $P = 120 \text{ W}$

Explanation: Solve the equation in question 60 for P.

62. $\dfrac{n_2}{n_1} = \dfrac{\lambda_1}{\lambda_2}$

$\lambda_2 = \dfrac{n_1 \lambda_1}{n_2}$

$\lambda_2 = \dfrac{1.00(5.89 \times 10^{-7} \text{m})}{1.47}$

Explanation: Under Waves in the RT, find the equation $n_2/n_1 = \lambda_1/\lambda_2$. The subscript of 1 refers to the original medium (air) and the subscript 2 refers to the new medium (corn oil). The values of n_1 and n_2 are found on the table of Absolute Indices of Refraction in the RT. Substitute the given values into the equation.

63. $\lambda_2 = 4.01 \times 10^{-7}$ m

Explanation: Solve the equation in question 62 for λ_2.

64. 4.22×10^{-2} u

Explanation: From the List of Physical Constants in the RT, 1 universal mass unit (u) = 9.31×10^2 MeV. Use this as a conversion factor to convert megaelectronvolts (MeV) to universal mass units:
39.3 MeV × (1 u)/(9.31×10^2 MeV) = 4.22×10^{-2} u.

65.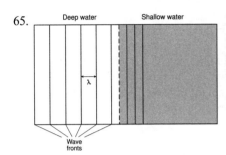

Explanation: Under Waves in the RT, find the equation $v = f\lambda$. Once a wave has been generated, its frequency remains constant. From this equation, if the speed decreases, the wavelength must decrease.

Part C

66. 1.0 cm = 0.20 m/s

Explanation: The vector representing the 1.50 m/s velocity is 7.5 cm long. One cm then represents 0.20 m/s: (1.50 m/s)/(7.5 cm) = (0.20 m/s)/cm.

67.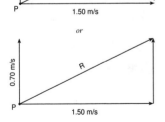

Explanation: The resultant velocity is the diagonal of the rectangle formed by the two component velocities when arranged tail to tail or the hypotenuse of the right triangle formed by the two component velocities when arranged head to tail.

68. 1.7 m/s

Explanation: Measure the length of the line R in either diagram and interpret according to the scale determined in 66. The length of R is 8.3 cm, which represents a velocity of 1.7 m/s (8.3 cm × 0.20 m/s).

69. 65° ± 2°

Explanation: Measure the angle between R and the 0.70 m/s velocity.

70. $F_s = kx$

$$x = \frac{F_s}{k} = \frac{mg}{k}$$

$$x = \frac{(2.00\,\text{kg})(9.81\,\text{m/s}^2)}{150.\,\text{N/m}}$$

Explanations: Under Mechanics in the RT, find the equations $F_s = kx$ and $g = F_g/m$. F_s is the weight (F_g) of the mass hanging from the spring. From the second equation, $F_g = mg$. The first equation can now be expressed as mg = kx. Solving for x, $x = mg/k$. The value of g is found on the List of Physical constants in the RT. Substitute the given values into the equation.

71. $x = 0.131$ m

Explanation: Solve the equation in question 70 for x.

72. $PE_s = \frac{1}{2}kx^2$

$$PE_s = \frac{1}{2}(150.\,\text{N/m})(0.131\,\text{m})^2$$

Explanation: In the RT under Mechanics, find the equation $PE_s = 1/2\,kx^2$. Substitute into the equation, using the value of x from question 71.

73. $PE_s = 1.29$ J

Explanation: Solve the equation in question 72.

74. 6.00 Ω

Explanation: Under Electricity in the RT, find the equation $R = V/I$. Substitution and solving gives $R = (12.0\,\text{V})/(2.00\,\text{A}) = 6.00\,\Omega$.

75. $R = \dfrac{\rho L}{A}$

$$A = \frac{\rho L}{R}$$

$$A = \frac{(150. \times 10^{-8}\,\Omega \bullet \text{m})(0.100\,\text{m})}{6.00\,\Omega}$$

Under Electricity in the RT, find the equation $R = \rho L/A$. The length must be in cm (10.0 cm = 0.1000 m). The resistivity of nichrome is found in the RT on the table of Resistivities at 20°C. Use the resistance value calculated in question 74.

76. $A = 2.50 \times 10^{-8}$ m²

Explanation: Solve the equation in question 75 for A.

77. $E_{photon} = \dfrac{hc}{\lambda}$

$$E_{photon} = \frac{(6.63 \times 10^{-34}\,\text{J}\bullet\text{s})(3.00 \times 10^8\,\text{m/s})}{2.29 \times 10^{-7}\,\text{m}}$$

Explanation: In the RT under Modern Physics, find the equation $E_{photon} = hc/\lambda$. The values of h and c are found on the List of Physical Constants in the RT.

78. $E_{photon} = 8.69 \times 10^{-19}$ J

Explanation: Solve the equation in question 77 for E_{photon}.

79. 5.43 eV

Explanation: In the List of Physical Constants in the RT, $1\text{ eV} = 1.60 \times 10^{-19}\text{ J}$. Use this as a conversion factor to convert J to eV: $8.69 \times 10^{-19}\text{ J} \times (1\text{ eV})/(1.60 \times 10^{-19}\text{ J}) = 5.43\text{ eV}$.

80. Yes, the photon can be absorbed.

Explanation: The energy of the photon is exactly equal to the energy-level difference between the ground state and the d level. In the RT under Modern Physics, find the equation $E_{photon} = E_i - E_f$. The values of the energy levels for mercury are found on the Energy Level Diagrams on the RT. The difference between the ground state and d level is $E_{photon} = (-10.38\text{ eV}) - (-4.95\text{ eV}) = -5.43\text{ eV}$. The negative sign indicates that the photon is absorbed by the atom.

81. $41° \pm 2°$

Explanation: The angle of incidence is the angle between the incident ray and the normal at the point of incidence.

82. $n_1 \sin\theta_1 = n_2 \sin\theta_2$ Explanation: Under Waves in the RT, find the equation

$$n_2 = \frac{n_1 \sin\theta_1}{\sin\theta_2}$$

$$n_2 = \frac{(1.00)(\sin 41°)}{\sin 20.°}$$

$n_1 \sin\theta_1 = n_2 \sin\theta_2$. The subscript 1 refers to air and the subscript 2 refers to that of the block. The index of refraction of air is found in the RT on the Absolute Indices of Refraction table. Use the angle of incidence measured in question 81.

83. $n_2 = 1.9$

Explanation: Solve the equation in question 82 for n_2.

84.

$$n = \frac{c}{v}$$

$$v = \frac{c}{n}$$

$$v = \frac{3.00 \times 10^8\text{ m/s}}{1.9}$$

or

$$\frac{n_2}{n_1} = \frac{v_1}{v_2}$$

$$v_2 = \frac{n_1 v_1}{n_2}$$

$$v_2 = \frac{(1.00)(3.00 \times 10^8\text{ m/s})}{1.9}$$

Explanation: Under Waves in the RT, find the equation $n = c/v$. The speed of light is found on the List of Physical Constants in the RT. Use the value of n for the block from question 83.

85. $v = 1.6 \times 10^8\text{ m/s}$

Explanation: Solve the equation in question 84 for v.

Helpful Hints for Physics

Physics is a science that uses many equations to arrive at a solution to a problem. To help you learn physics, here are a few helpful hints.

1. For a numerical problem, do your "bookkeeping". Read the problem through once. Now read it again and write down in list form the information given with the correct symbols and units along with the unknown. This is the "bookkeeping" part. Look at what you have written down and select the equation that relates to all of the variables. In some problems, two equations may have to be used. Drawing a labeled diagram can be especially helpful in problems such as those dealing with vectors or projectile (two dimensional) motion. Do this whenever you think a diagram will help you "see" the problem more clearly.

2. Use units in your equation setup and solution. Do the same algebra with the units as you do with the number in solving the problem. The correct unit for the unknown quantity should come from the units used in the setup. Arriving at an answer with the correct unit is a strong indication that your solution is correct. Dimensional analysis is a process of using units to solve a problem or make a conversion.

3. Know the physics reference table (RT) well. Know the type of information and its location in the RT. Many questions can be answered simply by looking in the right place in the RT. Commonly used equations along with the meaning of the symbols are in the RT, listed under the topic to which they refer.

4. Graphs are a convenient and useful way to show the relationship between variables in an equation. The following types occur frequently in physics:

    a) Direct relationship – mathematically, this is represented by $y = kx$, where k is a constant. In a simple direct relationship, as x increases, y increases at the same rate. If x doubles, y doubles. It is said that y varies directly with x. The graph is a straight line.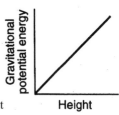

    Example: $\Delta PE = mg\Delta h$ where (mg) is constant

    b) Direct square relationship – mathematically, this is represented by $y = kx^2$, where k is a constant. In a direct square relationship, as x increases, y increases at a more rapid rate. If x doubles, y quadruples. It is said that y varies directly with the square of x. The graph is a curve.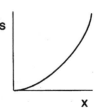

    Example: $PE_s = \frac{1}{2} kx^2$ where $(\frac{1}{2} k)$ is constant

c) Inverse relationship – mathematically, this is represented by $y = k/x$, where k is a constant. In a simple inverse relationship, as x increases, y decreases at the same rate. If x doubles, y halves. It is said that y varies inversely with x. The graph is a curve.

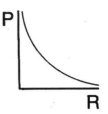

Example: $P = \dfrac{V}{R}$ where V is constant

d) Inverse square relationship – mathematically, this is represented by $y = \dfrac{k}{x^2}$, where k is a constant. In an inverse square relationship, as x increases, y decreases at a more rapid rate. If x doubles, y quarters. It is said that y varies inversely with the square of x. The graph is similar to that of the inverse relationship.

Example: $F_E = \dfrac{kq_1 q_2}{r^2}$ where $(kq_1 q_2)$ is constant

e) The slope of a graph is defined as the change in value of the y-coordinate divided by the change in value of the x-coordinate (commonly stated as rise over run): slope $= \dfrac{\Delta y}{\Delta x}$ where Δ means change.

Example: slope $= \dfrac{\Delta \text{work}}{\Delta \text{time}} = \dfrac{\text{joules}}{\text{seconds}} = $ watts

PHYSICAL SETTING
PHYSICS — REFERENCE TABLE
2006 EDITION

Contents

List of Physical Constants

Name	Symbol	Value
Universal gravitational constant	G	6.67×10^{-11} N•m^2/kg^2
Acceleration due to gravity	g	9.81 m/s^2
Speed of light in a vacuum	c	3.00×10^8 m/s
Speed of sound in air at STP		3.31×10^2 m/s
Mass of Earth		5.98×10^{24} kg
Mass of the Moon		7.35×10^{22} kg
Mean radius of Earth		6.37×10^6 m
Mean radius of the Moon		1.74×10^6 m
Mean distance—Earth to the Moon		3.84×10^8 m
Mean distance—Earth to the Sun		1.50×10^{11} m
Electrostatic constant	k	8.99×10^9 N•m^2/C^2
1 elementary charge	e	1.60×10^{-19} C
1 coulomb (C)		6.25×10^{18} elementary charges
1 electronvolt (eV)		1.60×10^{-19} J
Planck's constant	h	6.63×10^{-34} J•s
1 universal mass unit (u)		9.31×10^2 MeV
Rest mass of the electron	m_e	9.11×10^{-31} kg
Rest mass of the proton	m_p	1.67×10^{-27} kg
Rest mass of the neutron	m_n	1.67×10^{-27} kg

Prefixes for Powers of 10

Prefix	Symbol	Notation
tera	T	10^{12}
giga	G	10^9
mega	M	10^6
kilo	k	10^3
deci	d	10^{-1}
centi	c	10^{-2}
milli	m	10^{-3}
micro	μ	10^{-6}
nano	n	10^{-9}
pico	p	10^{-12}

Approximate Coefficients of Friction

	Kinetic	Static
Rubber on concrete (dry)	0.68	0.90
Rubber on concrete (wet)	0.58	
Rubber on asphalt (dry)	0.67	0.85
Rubber on asphalt (wet)	0.53	
Rubber on ice	0.15	
Waxed ski on snow	0.05	0.14
Wood on wood	0.30	0.42
Steel on steel	0.57	0.74
Copper on steel	0.36	0.53
Teflon on Teflon	0.04	

The Electromagnetic Spectrum

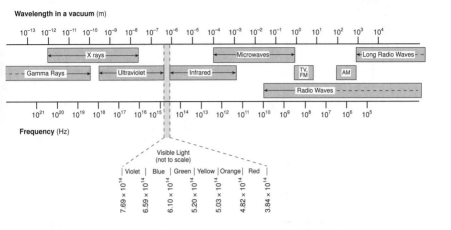

Absolute Indices of Refraction
$(f = 5.09 \times 10^{14} \text{ Hz})$

Air	1.00
Corn oil	1.47
Diamond	2.42
Ethyl alcohol	1.36
Glass, crown	1.52
Glass, flint	1.66
Glycerol	1.47
Lucite	1.50
Quartz, fused	1.46
Sodium chloride	1.54
Water	1.33
Zircon	1.92

Energy Level Diagrams

Hydrogen

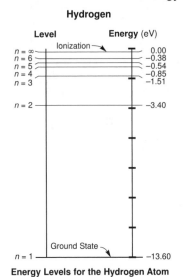

Energy Levels for the Hydrogen Atom

Mercury

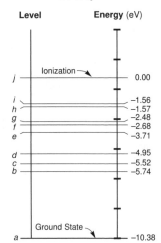

A Few Energy Levels for the Mercury Atom

Classification of Matter

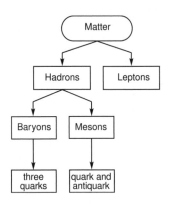

Particles of the Standard Model

Quarks

Name	up	charm	top
Symbol	u	c	t
Charge	$+\frac{2}{3}e$	$+\frac{2}{3}e$	$+\frac{2}{3}e$

	down	strange	bottom
	d	s	b
	$-\frac{1}{3}e$	$-\frac{1}{3}e$	$-\frac{1}{3}e$

Leptons

electron	muon	tau
e	μ	τ
$-1e$	$-1e$	$-1e$

electron neutrino	muon neutrino	tau neutrino
ν_e	ν_μ	ν_τ
0	0	0

Note: For each particle, there is a corresponding antiparticle with a charge opposite that of its associated particle.

Electricity

$$F_e = \frac{kq_1q_2}{r^2}$$

$$E = \frac{F_e}{q}$$

$$V = \frac{W}{q}$$

$$I = \frac{\Delta q}{t}$$

$$R = \frac{V}{I}$$

$$R = \frac{\rho L}{A}$$

$$P = VI = I^2R = \frac{V^2}{R}$$

$$W = Pt = VIt = I^2Rt = \frac{V^2t}{R}$$

A = cross-sectional area
E = electric field strength
F_e = electrostatic force
I = current
k = electrostatic constant
L = length of conductor
P = electrical power
q = charge
R = resistance
R_{eq} = equivalent resistance
r = distance between centers
t = time
V = potential difference
W = work (electrical energy)
Δ = change
ρ = resistivity

Series Circuits

$I = I_1 = I_2 = I_3 = \ldots$

$V = V_1 + V_2 + V_3 + \ldots$

$R_{eq} = R_1 + R_2 + R_3 + \ldots$

Parallel Circuits

$I = I_1 + I_2 + I_3 + \ldots$

$V = V_1 = V_2 = V_3 = \ldots$

$\dfrac{1}{R_{eq}} = \dfrac{1}{R_1} + \dfrac{1}{R_2} + \dfrac{1}{R_3} + \ldots$

Circuit Symbols

⊥ cell

⊥ battery

⟋ switch

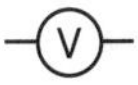

—(V)— voltmeter

—(A)— ammeter

/\/\/\ resistor

/\/\/\ variable resistor

—(lll)— lamp

Resistivities at 20°C	
Material	**Resistivity ($\Omega \cdot$m)**
Aluminum	2.82×10^{-8}
Copper	1.72×10^{-8}
Gold	2.44×10^{-8}
Nichrome	$150. \times 10^{-8}$
Silver	1.59×10^{-8}
Tungsten	5.60×10^{-8}

Waves

$v = f\lambda$

$T = \dfrac{1}{f}$

$\theta_i = \theta_r$

$n = \dfrac{c}{v}$

$n_1 \sin \theta_1 = n_2 \sin \theta_2$

$\dfrac{n_2}{n_1} = \dfrac{v_1}{v_2} = \dfrac{\lambda_1}{\lambda_2}$

c = speed of light in a vacuum
f = frequency
n = absolute index of refraction
T = period
v = velocity or speed
λ = wavelength
θ = angle
θ_i = angle of incidence
θ_r = angle of reflection

Modern Physics

$E_{photon} = hf = \dfrac{hc}{\lambda}$

$E_{photon} = E_i - E_f$

$E = mc^2$

c = speed of light in a vacuum
E = energy
f = frequency
h = Planck's constant
m = mass
λ = wavelength

Geometry and Trigonometry

Rectangle

$A = bh$

Triangle

$A = \frac{1}{2}bh$

Circle

$A = \pi r^2$

$C = 2\pi r$

A = area
b = base
C = circumference
h = height
r = radius

Right Triangle

$c^2 = a^2 + b^2$

$\sin \theta = \dfrac{a}{c}$

$\cos \theta = \dfrac{b}{c}$

$\tan \theta = \dfrac{a}{b}$

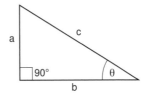

Mechanics

$$\bar{v} = \frac{d}{t}$$

$$a = \frac{\Delta v}{t}$$

$$v_f = v_i + at$$

$$d = v_i t + \frac{1}{2}at^2$$

$$v_f^2 = v_i^2 + 2ad$$

$$A_y = A \sin \theta$$

$$A_x = A \cos \theta$$

$$a = \frac{F_{net}}{m}$$

$$F_f = \mu F_N$$

$$F_g = \frac{Gm_1 m_2}{r^2}$$

$$g = \frac{F_g}{m}$$

$$p = mv$$

$$p_{before} = p_{after}$$

$$J = F_{net}\, t = \Delta p$$

$$F_s = kx$$

$$PE_s = \frac{1}{2}kx^2$$

$$F_c = ma_c$$

$$a_c = \frac{v^2}{r}$$

$$\Delta PE = mg\Delta h$$

$$KE = \frac{1}{2}mv^2$$

$$W = Fd = \Delta E_T$$

$$E_T = PE + KE + Q$$

$$P = \frac{W}{t} = \frac{Fd}{t} = F\bar{v}$$

a = acceleration

a_c = centripetal acceleration

A = any vector quantity

d = displacement or distance

E_T = total energy

F = force

F_c = centripetal force

F_f = force of friction

F_g = weight or force due to gravity

F_N = normal force

F_{net} = net force

F_s = force on a spring

g = acceleration due to gravity or gravitational field strength

G = universal gravitational constant

h = height

J = impulse

k = spring constant

KE = kinetic energy

m = mass

p = momentum

P = power

PE = potential energy

PE_s = potential energy stored in a spring

Q = internal energy

r = radius or distance between centers

t = time interval

v = velocity or speed

$\bar{v}$ = average velocity or average speed

W = work

x = change in spring length from the equilibrium position

Δ = change

θ = angle

μ = coefficient of friction

Correlation of Questions to Topics

<u>June 2008-</u>

<u>Mechanics</u> – 1, 2, 3, 4, 5, 6, 7, 8, 9, 10, 11, 12, 13, 14, 15, 16, 17, 18, 19, 36, 37, 38, 39, 40, 41, 42, 43, 44, 45, 52, 53, 54, 55, 56, 57, 62, 63, 64, 68, 70, 71

<u>Waves and Optics</u> – 25, 26, 27, 28, 29, 30, 31, 32, 48, 61, 72, 73, 74

<u>Electricity</u> – 20, 21, 22, 23, 24, 46, 47, 58, 59, 60, 65, 66, 67

<u>Modern Physics</u> – 33, 34, 35, 49, 50, 51, 75, 76

<u>Graphing</u> – 42, 45, 49, 68, 69

<u>June 2009-</u>

<u>Mechanics</u> – 1, 2, 3, 4, 5, 6, 7, 8, 9, 10, 11, 12, 13, 14, 15, 16, 17, 34, 36, 37, 38, 39, 40, 41, 42, 43, 48, 49, 50, 51, 51, 60, 61, 62, 63, 64, 65

<u>Waves and Optics</u> – 23, 24, 25, 26, 27, 28, 29, 30, 33, 57, 58, 59, 66, 67, 68, 69, 70

<u>Electricity</u> – 17, 18, 19, 20, 21, 22, 45, 46, 47, 53, 54, 55, 56

<u>Modern Physics</u> – 31, 32, 35, 44, 70, 71, 72

<u>Graphing</u> – 45

<u>June 2010-</u>

<u>Mechanics</u> – 1, 2, 3, 4, 5, 6, 7, 8, 9, 10, 11, 12, 13, 14, 15, 16, 17, 34, 36, 38, 39, 40, 42, 43, 44, 47, 51, 52, 53, 54, 56, 59, 60, 64, 65, 66, 67, 68

<u>Waves</u> – 24, 25, 26, 27, 28, 29, 30, 35, 48, 49, 50, 69, 70, 71, 75

<u>Electricity</u> – 18, 19, 20, 21, 22, 23, 37, 41, 45, 46, 47, 55, 57, 58

<u>Modern Physics</u> – 31, 32, 33, 34, 72, 73, 74

<u>Graphing</u> – 61, 62, 63

<u>June 2011-</u>

<u>Mechanics</u> – 1, 2, 3, 4, 5, 6, 7, 8, 9, 10, 11, 12 13, 14, 36, 37, 38, 39, 41, 42, 51-52, 53, 54-55, 56-57, 59, 66, 67, 68, 69, 70-71, 72-73

<u>Waves</u> – 16, 25, 26, 27, 28, 29, 30, 31, 32, 33, 46, 47, 48, 50, 62-63, 65, 81, 82-83, 84-85

<u>Electricity</u> – 15, 17, 18, 19, 20, 21, 22, 23, 24, 40, 43, 44, 45, 58, 60-61, 74, 75-76

<u>Modern Physics</u> – 34, 35, 49, 64, 77-78, 79, 80

<u>Graphing</u> – 37, 41, 42, 43, 59